POWER SYSTEM CONTROL AND PROTECTION

Academic Press Rapid Manuscript Reproduction

POWER SYSTEM CONTROL AND PROTECTION

EDITED BY

B. Don Russell

Electric Power Institute
Department of Electrical Engineering
Texas A&M University
College Station, Texas

Marion E. Council

School of Electrical Engineering and Computing Sciences
University of Oklahoma
Norman, Oklahoma

ACADEMIC PRESS   New York San Francisco London   1978
A Subsidiary of Harcourt Brace Jovanovich, Publishers

Copyright © 1978, by Academic Press, Inc.
ALL RIGHTS RESERVED.
NO PART OF THIS PUBLICATION MAY BE REPRODUCED OR
TRANSMITTED IN ANY FORM OR BY ANY MEANS, ELECTRONIC
OR MECHANICAL, INCLUDING PHOTOCOPY, RECORDING, OR ANY
INFORMATION STORAGE AND RETRIEVAL SYSTEM, WITHOUT
PERMISSION IN WRITING FROM THE PUBLISHER.

ACADEMIC PRESS, INC.
111 Fifth Avenue, New York, New York 10003

United Kingdom Edition published by
ACADEMIC PRESS, INC. (LONDON) LTD.
24/28 Oval Road, London NW1 7DX

Library of Congress Cataloging in Publications Data

Main entry under title:

Power system control and protection.

    1. Electric power systems—Automation—Addresses, essays, lectures. 2. Electric power systems—Protection—Addresses, essays, lectures.
I. Russell, B. Don. II. Council, Marion E.
TK1005.P717   621.31'7   78-6653
ISBN 0-12-604350-7

PRINTED IN THE UNITED STATES OF AMERICA

CONTENTS

Contributors vii
Preface ix

Functional and Design Considerations
for an Intelligent Remote Terminal 1
    G. Brant Laycock and Donald F. Koenig

Implementation of the SCADA Control
Center Functions 23
    Max D. Anderson

Load Management—A Systems Concept 63
    Julius Orban

The Development and Selection of Algorithms
for Relaying of Transmission Lines
by Digital Computer 83
    J. G. Gilbert, E. A. Udren, and M. Sackin

Fault Current Limiters 127
    Harry J. King and Richard E. Kennon

Hybrid Simulation of Electric Power Systems 167
    Lewis N. Walker

Elements of Power System Identification
and Control 197
    Bruce K. Colburn

Synchronous Machine Modeling: A System Comparison 223
    John E. Fagan

Control of Wind Turbine Generators Connected
to Power Systems 239
    H. H. Hwang, H. V. Mozeico, and L. J. Gilbert

Operation and Control of Wind–Electric Systems 261
    R. Ramakumar

CONTRIBUTORS

Numbers in parentheses indicate pages where authors' contributions begin.

MAX D. ANDERSON (23), Electrical Engineering Department, University of Missouri–Rolla, Rolla, Missouri

BRUCE K. COLBURN (197), Department of Electrical Engineering, Texas A&M University, College Station, Texas

JOHN E. FAGAN (223), School of Electrical Engineering, The University of Oklahoma, Norman, Oklahoma

J. G. GILBERT (83), Pennsylvania Power and Light Company, Allentown, Pennsylvania

L. J. GILBERT (239), NASA, Lewis Research Center, Cleveland, Ohio

H. H. HWANG (239), Department of Electrical Engineering, University of Hawaii, Honolulu, Hawaii

RICHARD E. KENNON (127), Electric Power Research Institute, Palo Alto, California

HARRY J. KING (127), Hughes Research Laboratories, Malibu, California

DONALD F. KOENIG (1), TRW Controls, Houston, Texas

G. BRANT LAYCOCK (1), TRW Controls, Houston, Texas

H. V. MOZEICO (239), Department of Electrical Engineering, University of Hawaii, Honolulu, Hawaii

JULIUS ORBAN (63), Sangamo Electric Company, Springfield, Illinois

R. RAMAKUMAR (261), School of Electrical Engineering, Stillwater, Oklahoma

M. SACKIN (83), Westinghouse Electric Corporation, Churchill, Pennsylvania

E. A. UDREN (83), Westinghouse Electric Corporation, Newark, New Jersey

LEWIS N. WALKER (167), Electrical Engineering Department, University of Missouri–Columbia, Columbia, Missouri

PREFACE

Our society depends greatly on energy in the form of electricity. As this need grows, the control and protection of power systems to ensure a reliable and secure supply become even more critical. It is the job and responsibility of power systems engineers to apply the latest technology to see that this goal is met.

The problems surrounding the generation, transmission, distribution, and utilization of electric power seem overwhelming when viewed as a whole. However, our traditional solution to difficult problems works very well. Through study, dedication, and experience, individual engineers become experts in specific technological areas. They then apply their skills to a particular problem until a solution is obtained. It is to these engineers that this volume is dedicated. We have selected a group of subjects that are representative of today's developing technologies as applied to power systems. Individual authors were selected who are recognized experts in each of these areas. They know the problems and, guided by practicality, have found answers. The contents of this volume may seem diverse, but so are the problems that beset us. Each power systems engineer should find something to which he strongly relates.

We include papers on intelligent remote terminals and SCADA control functions because remote control of substations and system data acquisition are critical to modern operations philosophy. Load management is of great importance considering supply problems we shall experience over the next generation. Papers on computer relaying and fault current limiters represent the forefront of modern system protection. The subjects of hybrid simulation, system identification, and modeling have been included. They represent the cross fertilization of state-of-the-art techniques from other disciplines applied to power system problems. Finally, papers on wind generation are included as representative of the new technologies we face as we seek additional supplies of energy.

We are pleased to acknowledge the assistance of Academic Press in the preparation of the volume. Also, for their encouragement, we thank Dr. W. B. Jones, Head, Department of Electrical Engineering, Texas A&M

University, and Dr. C. R. Haden, Director, School of Electrical Engineering and Computing Sciences, University of Oklahoma.

The papers in this volume represent advances in power system technology. We hope they serve as motivation for additional work that will further advance power system control and protection.

Power Control System and Protection

FUNCTIONAL AND DESIGN CONSIDERATIONS FOR AN INTELLIGENT REMOTE TERMINAL

G. Brant Laycock and Donald F. Koenig

TRW Controls Corporation

I. INTRODUCTION

The capabilities and limitations of intelligent remote terminal units are explored and discussed in this chapter. Functional and design features that have been, are being, and could be implemented in an intelligent unit are defined and evaluated.

A remote terminal is a part of a process and/or supervisory control and data acquisition system (SCADA) in which automatic functions are implemented within the remote terminal. An intelligent remote (See Figure 1) is one in which much of the logic associated with each function has been programmed and is executed in a digital processor.

The environmental and functional requirements for the remote are defined and discussed. Standard functions and capabilities (those traditionally associated with data acquisition and supervisory control) are considered mandatory. Additional functions of a computational and/or automatic control nature are candidates for inclusion in the intelligent remote, particularly when they use data already available, or offer operational advantages to the user. Each candidate function is defined, and its operating requirements relative to the capabilities of the intelligent remote are evaluated.

The operational timeline and resource utilization in the intelligent remote is discussed in detail. The impact of a low <u>vs</u> high performance processor on the implementation of capabilities in an intelligent remote is illustrated. The future functional expandability requirements for this remote are discussed. References have been included to aid in defining the environmental requirements and to illustrate state-of-the-art activities that involve intelligent remote terminal units and their capabilities.

II. OPERATING ENVIRONMENTS

The remote terminal unit may be required to operate and be maintained in a wide variety of environments. Reference 1 defines typical environmental and maintenance requirements

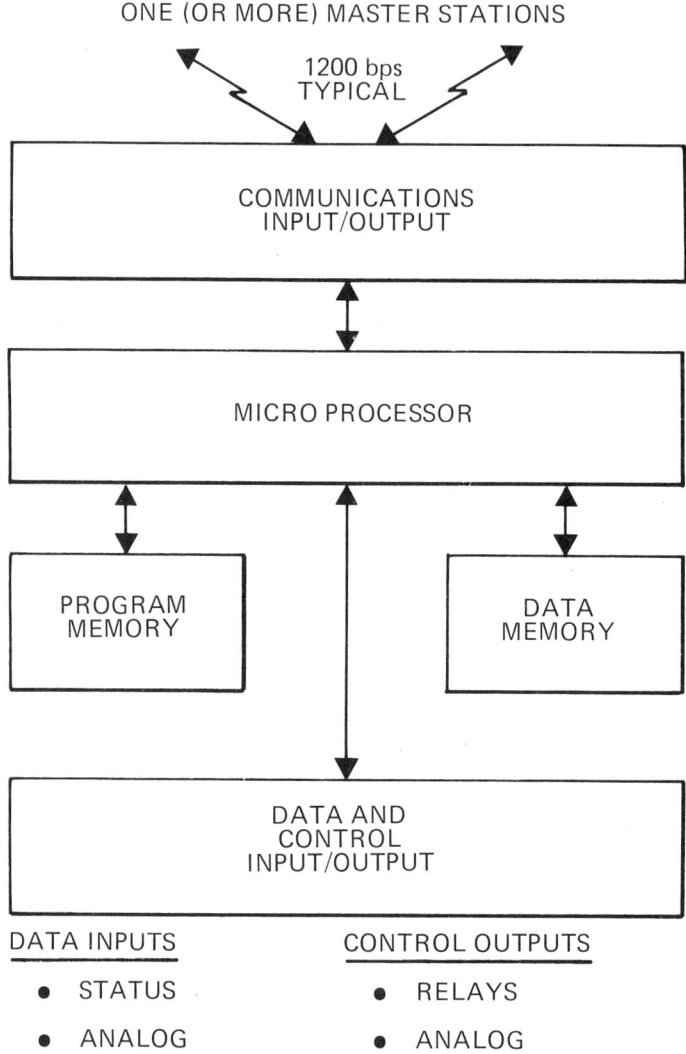

FIGURE 1
INTELLIGENT REMOTE TERMINAL UNIT

applicable to the remote terminal. The environmental requirements from reference 1 and supporting information on possible Electromagnetic Interference (EMI) of certain environments (reference 2) are the basis for the requirements (See Table 1) applicable to the intelligent remote terminal unit.

The environmentally related requirements in Table 1 summarize some of the factors influencing the selection of a microprocessor and the design of intelligent remote terminal units. Few off-the-shelf microprocessors can satisfy the temperature and EMI environmental requirements; therefore, special provisions or a special microprocessor must be provided. The environmental requirements are applicable to the mandatory and candidate functions discussed below.

A second level of environmental considerations exists in the design and application of intelligent remotes. In what logic environment is the digital processor expected to operate? The environment for the processor must provide not only the data and instructions associated with a particular function but, more importantly, it must provide these inputs in a predefined priority order. The priority assigned to a function is of course related to the importance of that function to the successful and proper operation of the intelligent remote unit as a part of a system. The controlling logic in the processor must decide how and when the processing resources are to be applied. The controlling logic therefore requires interfaces and interface protocol in order to properly allocate the processing resources to the highest priority function. The methods used by the controlling logic to manage processing resources include the following:

    Event Interrupts - A hardware contact is closed

    Time Interrupts -- A tick of the clock has occurred

    Timer Interrupts - A predefined number of ticks has occurred

    Flag ------------- A predefined condition is recognized.

The use of one or more of the above methods for a particular function within the intelligent terminal unit has an impact on unit performance as well as on unit cost. For example, a high performance processor that has time to poll for and use flags (e.g., has the status changed?) might be a less costly alternative than the inclusion of the necessary logic to generate event interrupts (e.g., status has changed) on every status input card. These implementation alternatives are discussed further in a later section after the functions have been defined and the criteria for resource utilization has been discussed.

Table 1   Summary of Environmental Requirements

Temperature:	-25° to +60° Celsius
Maximum Rate of Change:	20° Celsius per hour
Chemical:	Site specific
Altitude:	Up to 2000 meters
EMI Radiated Fields:	0 to 1 volt/meter/megahertz
EMI Conducted:	IEEE Surge Withstand Criteria (SWC) (IEEE Std. 472-1974)
Seismic, Shock and Vibration:	Site specific plus maintenance handling
Lightning Protection:	Site specific
Acoustic Energies:	Limits are specified in Ref. 1
Maintainability:	As required in concert with reliability to satisfy system availability criteria
Security of Operation:	Probability of undetected bit error, impact of failures, select/checkback before operate, function checks, alarm criteria are defined in Ref. 1
Source Voltages:	4 ranges of dc, 2 ranges of ac
Communications:	Common mode SWC (IEEE Std. 472-1974)
a) Remote to Master: (Non-integral modem)	EIA Stds., RS-232-C, 363 & 404 ANSI Stds., X3.1-1969, X3.24-1969
b) Remote to Master: (Integral modem)	Electrical and protocol standards, channel utilization calculations. Bit serial communications rate may be as low as 300 bps or as high as 1 Mega bps depending upon the application and communication media, associated with geographically distributed or in-plant systems.
Input/Output:	Analog, digital electronic, and digital electromechanical points as required by the application.

III. HARDWARE VS FIRMWARE ALLOCATION CRITERIA

The intelligent remote is a combination of analog and digital circuits where the digital circuits may be further divided between special purpose and general purpose (e.g., those that make up the microprocessor). During the design phase of the intelligent remote an allocation of the mandatory and candidate functions is made to special and general purpose circuits. Those functions (or portions of functions) that are allocated to the general purpose digital circuitry will be implemented in firmware.

The criteria and guidelines used in assigning various functions to hardware or firmware are discussed in this section. These criteria have a major impact on the architectural and performance requirements of the microprocessor used in the intelligent remote. The development and application of such criteria is an iterative process which may take many months.

Three design criteria are defined and used in hardware/firmware trade-offs to illustrate the allocation process. The three design goals are:

o Reduce the number of unique function cards

o Provide master/remote communications flexibility

o Minimize the microprocessor executive overhead.

Unique Function Cards - Earlier generation remote units have many unique circuit cards, each implementing a different function. The best example of this are the various types of status (discrete) inputs.

o Normal - value (1 or 0) whenever read

o Latching - value set to 1 if contacts close

o Memory - 2 state, 3 state, requires multiple bits

o Pulse Counting - the number of 0 to 1 transitions

o Pulse Duration - the time duration of a contact closure.

When the designs for the individual status function cards are overlayed the signal interface circuits that filter, condition and recognize when the external status contacts are closed are usually identical. The circuitry on each type of card that uses the conditioned signal gives one card a personality different from its neighbor. Defining the personality logic in equation form is a necessary step in determining what must be done in firmware in order to use a single discrete input card design to satisfy the different status input requirements. The equations will define what the firmware must do. We must, however, look again at the personality of the individual point interface in order to determine how often the firmware is to be exercised. The discrete inputs must be services as follows:

- o Normal - read upon command
- o Latching - read to satisfy latch requirements
- o Memory - read to satisfy specific criteria
- o Pulse Counting - read to recognize transitions
- o Pulse Duration - read to satisfy resolution requirements.

The cyclic read requirements associated with the various status input types depend primarily upon the design of the point interface circuitry. The firmware used with one discrete input card design is executed every 12 milliseconds (reference 3) to satisfy the requirements imposed by that application. The number of instruction executions depend upon the number of logical steps used to implement each type of status input and the maximum number of status inputs to be serviced. To illustrate the impact status personality processing has on the performance requirements of the microprocessor, the following remote terminal requirements are established.

- o A maximum of 480 status inputs are assumed (at 16 per card = 30 cards)
- o A maximum of 10% processor loading for status processing (scan every 12 ms = 1.2 ms per scan)
- o An average of 20 instructions per status personality is assumed (30 cards x 20 instructions = 600 instructions per scan)

These implementation oriented requirements indicate that the microprocessor must complete 600 instructions in 1.2 milliseconds or have an instruction time of 2 microseconds. If the average logical instruction is 2 memory cycles then the microprocessor must have a memory cycle of 1 microsecond.

<u>Master/Remote Communications</u> - To define the flexibility that may be required for the master/remote communications interface, one must study the various message standards (see Figure 2) the intelligent remote is expected to be able to accommodate.

MESSAGE ESTABLISHMENT	INFORMATION/COMMANDS	MESSAGE TERMINATION

Fig. 2   Typical Master/Remote Message Standard

If the fields within the three major segments of a message
standard fall on byte (5, 6, 7 or 8 bits) boundaries then a
hardware interface that buffers a byte at a time could be
used. Unfortunately many message standards in use today are
bit oriented, therefore, a bit buffer interface must, for
maximum flexibility, be accommodated. Another consideration
related to message standards is accommodating various error
detection codes. For maximum communications security these
codes are selected based upon message length and vary from
system to system. Encoding and decoding the error detection
code becomes, therefore, another candidate for implementation
in firmware. Computation of the error code can be accomplished while the communications channel is inactive, however,
it must be accomplished quickly in order to minimize the
message turnaround time for valid master-to-remote messages.
Communications flexibility when viewed at the unit level may
be summarized in the following remote terminal unit requirements.

o   Buffer one bit at a time

o   Accommodate channels up to 1200 bps
    (bit time = .833 milliseconds)

o   Maximum of 30% processor loading
    (service interface in less than .250 milliseconds)

Status personality processing established the microprocessor
speed at 2 microseconds per instruction, therefore, approximately 125 instruction times could be used to service this
interface. The intelligent remote analyzed in reference 3
uses less than 100 instructions to service the typical interrupt from the communications interface. The total size of
the service routine is approximately 300 instructions.

<u>Executive Overhead</u> - The communications interface is one
of several interfaces where microprocessor executive overhead may become burdensome. This interface may be serviced
via polling (is a bit present?) or the communications interface hardware may raise an interrupt when a bit time (transmit or receive) has elapsed. An interface that generates an
interrupt for each transmit or receive bit is preferred (ref.
4) and has been implemented (ref. 3). Implementing the communications interface as an interrupting interface requires
that the microprocessor be able to handle external hardware
interrupts in addition to the real-time clock interrupts.
The advantage of the interrupting interface over the polling
approach is a reduction of executive overhead since the
communications interface to an intelligent remote is dormant
most of the time. However, while active, this interface
could require service every bit time for hundreds of milliseconds, depending upon message length and channel speed.

The executive overhead may be further minimized in some architectures with a hybrid approach of polling this interface when messages (transmit or receive) are present and using the initial interrupt to schedule such a cyclic poll. Executive overhead and its impact depends to a great extent upon the design and capabilities of the microprocessor. A microprocessor with both real-time clock and external interrupt is considered mandatory.

Features of the microprocessor design also have an impact on the executive overhead. Separation of data and instruction bus lines may both reduce overhead and enhance throughput. Features of the microprocessor and its program and data memories are beyond the scope of this discussion.

IV. MANDATORY FUNCTIONS AND CAPABILITIES

Standard functions and capabilities of a SCADA remote are defined and discussed in references 1 through 4. The mandatory functions and capabilities are summarized in Table 2.

Table 2   Mandatory Functions and Capabilities - SCADA Remote

  o DATA ACQUISITION

     - Discrete Inputs

        Status Monitoring
        Pulse Counting
        Pulse Timing

     - Digital Inputs

        Byte (e.g., ASCII) Codes
        Binary Coded Decimal

     - Analog Inputs

        Single Ended (Common Return)
        Differential

  o CONTROL OUTPUTS

     - Supervisory (Manually Initiated)

        Checkback-Before-Operate

     - Automatic (Software or Firmware Initiated)

        Immediate Operate

CONSIDERATIONS FOR A REMOTE TERMINAL

- o MASTER/REMOTE COMMUNICATIONS
    - Universal Commands (All Remotes)
    - Message Standard
    - Line Protocol
- o OPERATIONAL SECURITY
    - Minimum of False Commands
        - Communications Errors
        - Component Failure
        - Electromagnetic Interference
    - Internal Monitoring
        - Power Failure/Restart
        - Analog Reference Inputs
        - Excitation Supply Inputs
- o FIELD EXPANDABILITY
    - Pre-Wired Spare
        - Plug-In Modules
    - Space-Only
        - Wiring and Modules
- o AVAILABILITY OF FUNCTIONS
    - Long Mean Time Between Failures (MTBF)
    - Short Mean Time to Restore (MTTR)

The intelligent remote terminal must be able to perform the above functions as efficiently as its hard-wired counterpart and have sufficient resources available to implement some of the additional functions defined in the next section. The mandatory functions and capabilities require that the design of the intelligent remote accommodate the following:

- o DATA ACQUISITION VARIATIONS
    - Point Counts (0 to some maximum)
    - Point Types (Analog, digital, discrete)
    - Scan Rates (Application dependent)
    - Buffering (e.g., accumulator freeze)
- o FIELD EXPANSION (WITHOUT NEW FIRMWARE)
    - Add/Delete Point Types
- o RAPID RESTORATION OF FUNCTION
    - Ease of Fault Isolation
    - Ease of Module Replacement

An intelligent unit that is capable of self-initialization (reference 3) can easily satisfy the point count variations that exist between different installation locations. Scan rates and point types are typically the same for all remotes of a system; therefore, these instructions may be imbedded in the firmware. Fault isolation may utilize diagnostic instructions that are executed in response to a technician's manual action.

Possibly the most important design consideration relative to the intelligent remote is how to executively control the resources of the microprocessor such that data are not lost and are processed correctly. Table 3 lists the mandatory functions and capabilities in their decreasing order of priority. The ordering is appropriate for an interrupt driven real-time digital processor that is assumed to be the controlling element of the intelligent remote terminal unit. The two top priority functions are not expected to occur very often. The real-time clock interrupt will occur at a cycle interval of 1 millisecond (or less) as required by the function with the highest resolution. The communications related functions (service, interpret, execute, transmit) occur in response to a message from the master station, typically every 2 to 60 seconds. These communications functions will use processor resources that are used for local data acquisition or special functions. If the communications related budget of 30% processor utilization is met, then some of the lower priority functions will be accomplished while the communications related functions are active.

Table 3 Priority Ordering of Mandatory Functions and Capabilities

Function	Remarks
Safe Powerfail Shutdown	Preclude erroneous operation due to low input power.
Internal Error Traps	Preclude erroneous operation due to internal errors.
Increment Clocks & Timers	Maintain accurate time bases for cyclic tasks.
Service Communications Interface	Prevent the loss of data.
Interpret Received Data	Establish priority of next action.
Execute Commands	Minimize delay for immediate operate commands.
Transmit Data	Answer MASTER STATION requests.

CONSIDERATIONS FOR A REMOTE TERMINAL

Acquire Status Data	Satisfy status point personality requirements.
Acquire Analog Data	Maintain a fresh internal data base.
Execute Diagnostics and Time-Outs	Reset interfaces as required.

V. ADDITIONAL FUNCTIONS AND CAPABILITIES

In this section additional requirements are defined as a basis for the later discussion of resource utilization in the intelligent remote unit.

Table 4 contains definitions of functions and capabilities that are implemented and/or are within the state-of-the-art for implementation (see references 5 through 8) in intelligent remote units.

Table 4   Definitions of Candidate Functions

Report By Exception o Status o Analog	The recognition that instrumented point values have changed; recording all changes for each point until a report to the master station is transmitted; maintaining that information until its error free reception is acknowledged; accepting change threshold criteria for analog points (e.g., ±3 bits out of the 12 bit digital representation).
Data Concentration	The suppression of data that are unchanging in such a manner that the eventual data user can reconstruct the time profile of instrumented quantities. Simple (e.g., magnitude constant) or complex (e.g., magnitude, frequency and phase angle all constant) decision criteria may be applied in intelligent remotes for various forms of data concentration.
Sequence of Events Reporting	The recognition that predefined event(s) have occurred; time tagging each event; reporting (locally and/or at the master station) the event data (e.g., Breaker T37 TRIP 13:47:37:352) in the order of occurrence of the events; accepting time synchronization signals from an

Sequence of Events Reporting (Cont'd.)	external source such that events from multiple remotes have a known system level time resolution (e.g., 8 milliseconds) maintaining a local event time resolution for events (e.g., 4 milliseconds).
Digital Metering	The computation of measurable quantities such as watts, vars, phase angle, fluid flow, etc. from their fundamental ingredients such as current, voltage, time, pressures, etc.
Fault Data Capture	The capture in digital form of pre-fault data (e.g., one cycle) and post fault data (until the fault is cleared, e.g., 3 to 10 cycles) with sufficient resolution to support digital relaying and/or fault reporting.
Fault Reporting	The processing of fault data and the output of a report in a tabular format. Information such as fault duration, max over currents, fault type (e.g., AØ to CØ, 3Ø to G), and estimated fault location are representative of the derived data.
Digital Relaying	The processing of fault data in order to decide where to clear the fault; accepting transfer trip/blocking commands as appropriate, all within the time (e.g., 1 or 2 cycles) allocated for the decision subfunction.
Control Sequences	The execution of a predefined sequence of stored commands in response to a single request.
Automatic Retry	The repetition of a data request or command for a predefined number of items (e.g., multiple shot reclosure).
Closed Loop Control o Voltage Level o Var Level o Interlocks o Synchronization	The application of stored criteria and/or algorithms to instrumented data for the automatic control of external apparatus.

TABLE 5: FUNCTIONAL REQUIREMENTS

FUNCTION	RESOLUTION	FREQUENCY	REMARKS
REPORTS BY EXCEPTION			
o Status	16 msec	Random	Up to 512 points
o Analog	500 msec	Random	Up to 256 points
DATA CONCENTRATION	Application specific	Cyclic, to satisfy resolution	Up to 64 points
SEQUENCE OF EVENTS	1 to 4 msec	Random	Up to 128 points
DIGITAL METERING	12 bits/value	1 sec	Up to 32 points
FAULT DATA CAPTURE	12 bits/value	1.33 msec	Up to 64 points 12 prefault samples/point
FAULT REPORTING	Best possible from Captured Fault Data	Random	Several per day to one in several years
DIGITAL RELAYING	Decision in 16-32 ms	Random	To clear fault condition
CONTROL SEQUENCES	Exactly as stored	Random to cyclic	Depends upon application
AUTOMATIC RETRY	Exactly as defined	Random, X times each	X defined for each type of controlled point
CLOSED LOOP CONTROL	Exactly as defined	Random to cyclic	Application dependent, external decision criteria.

The intelligent remote may contain one or more of the above functions in addition to the mandatory functions and capabilities discussed in the previous section.

The functional requirements for each of the candidate functions defined in Table 4 are summarized below in Table 5 in order to assess the impact of their implementation in the intelligent remote unit.

The inclusion of one or more of the additional functions with the mandatory functions and capabilities in an intelligent remote requires careful evaluation of worst case loading. What combination of events could occur (e.g., two simultaneous faults) that would compromise one or more functions? The remote will respond to the highest priority (most important) functional requirement before performing a lower priority function. In Table 6 selected functions have been priority ordered to illustrate the trade-offs facing the designer.

Table 6   Typical Priority Ordering of Candidate Functions

FUNCTION	PRIORITY	REMARKS
Fault Data Capture	1	Input to Digital Relaying, Pre-and Post-Fault Data
Digital Relaying	2	Decision processing overlaps Fault Data Capture
Sequence of Events (Data Capture)	3	SOE Data Capture must be able to overlap Fault Data Capture
Report By Exception (Data Capture)	5	To maintain required resolution
Closed Loop Control	8	May be suspended if Digital Relaying is Active
Digital Metering	10	May skip parameters associated with faulted apparatus
Fault Reporting	12	Reports for human usage may be output seconds after clearance of the fault.
Sequence of Events (Reporting)	14	

The priority one function from Table 6, "Fault Data Capture" must be accomplished every 1.33 msec. in order to capture 12 analog samples per cycle (60 Hz) for each point. These data, stored in a circular file, will be used as pre-fault reference data in the generation of the fault report. If 64 analog channels were instrumented 768 samples would be required for one cycle (16.7 msec.) of historical data. In the event that a fault condition occurs at least a subset of the 64 channels must be preserved as part of a fault data file while the 1 cycle history file is maintained for all normal signals.

Recognition of a suspected fault condition might be an output of the Digital Relaying function. Other outputs of the Digital Relaying function may be events and associated times of occurrence for the Sequence of Events Function. It is the intersection of these high priority functions that must be carefully analyzed in defining the intelligent remote units functional mix of capabilities.

The method of implementation for each function is a major input to the detailed analysis. For example, a fault data capture subsystem that placed the digitized representations of the analog quantities in memory and then signaled the executive control logic would place an insignificant burden on the processing logic. A preliminary processing of the captured data by the fault data capture subsystem could be used to initiate the Digital Relaying function whenever the captured data deviated from its sinusoidal pattern by more than a fixed amount in any sample time. The point of this example of a hardwired change detector is to illustrate that high priority functions do not necessarily utilize processor resources that are allocated to routine and repetitive functions. One must, however, analyze what happens when say the Digital Relaying Module is required to process multiple samples of fault data over a 100 millisecond interval, and issue fault clearance commands during that time period.

VI. RESOURCE PLANNING AND UTILIZATION

The previous sections discussed how the performance of an intelligent remote may be impacted by the many considerations such as point count, processor speed and flexibility, processing functions, communication requirements, time criticality of data, and I/O card complexity. The design goal for implementing an intelligent remote is to produce a remote which performs the mandatory functions defined in Section IV and still have adequate resources available in terms of processor time and memory to perform selected additional functions such as those discussed in Section V. The activities internal to the remote are directly related to the number and type of I/O cards in the remote and the functions to be performed. The following card types and point counts per card are characteristic of the type devices which may be interfaced to the remote.

Discrete input card	16 points
Analog input card	16 points
Control output card	16 relays
Matrix control output card	128 relays

These card types may be used in various combinations to configure a remote. Each input/output function requires a certain percentage of the remote's processor resources. For a specific configuration, the base processor loading may be established. The bit rate for the communication line also demands a level of processor loading since the volume of information to be handled by the message processing routines is controlled by the bit rate.

Assume a "typical" remote (details of this "typical" remote are presented in Reference 3) to be configured as having 256 momentary status points, 256 status with memory points, 96 analog inputs, 16 accumulator points, and 512 relay outputs. In addition, a 2400 bit per second communication interface is assumed which uses bit buffering as discussed in Section III. The processor is a high performance microprocessor with cycle times near 220 nanoseconds and a basic 16 bit architecture.

The processor scans the discrete input cards each 12 milliseconds to guarantee detecting status changes since the discrete input cards provide a 16 ±4 millisecond filtering of the status inputs. When active, the communication interface provides a data bit each 417 microseconds for the 2400 bit per second channel for firmware processing. These two functions are the primary users of processor time. The analog in- inputs require very little processor time since only one point is processed each two milliseconds. A one millisecond clock interrupt must also be accommodated.

The priority of processing by the firmware is divided between interrupt driven events and scheduled events. The processing of scheduled events is temporarily suspended when an interrupt event occurs. The clock interrupt is the highest priority interrupt event followed by the communication interface interrupts. A scheduled event will require running to completion before another scheduled event may start. The scheduled events are prioritized as status input processing followed by analog input processing. These four processing functions (real time clock and communications interrupts, status and analog input processing) may be used to establish the peak loading of the microprocessor in the remote for the mandatory SCADA functions.

Supervisory control operations do not take place during communication activity so the control timing impact need not be included in this analysis. The resource utilization of command processing for supervisory control is significantly less than that associated with communications I/O, therefore adequate resources exist to exercise controls after communication I/O is completed.

CONSIDERATIONS FOR A REMOTE TERMINAL

The status processing load is a linear function with respect to the number of discrete input cards. The communication interface processor load is also basically linear with respect to the communication bit rate. The processor load for the communication interface may be reduced by a factor of more than five if a byte rather than bit interface card is utilized for byte oriented message protocols.

For this typical remote configuration, the peak processor utilization is presented in Table 7. The table includes the frequency of execution, peak processing time per execution cycle, total execution time for a 12 millisecond interval, and percent processor utilization.

Table 7   Elements of Timing Analysis

FUNCTION	EXECUTION FREQUENCY	PEAK EXECUTION TIME	TOTAL EXECUTION TIME (12 ms)	PER CENT UTILIZATION
CLOCK	1 ms	33 µs	396µs	3.3%
COMMUNICATIONS	.417 ms	48µs	1392µs	11.6%
STATUS INPUTS	12 ms	788µs	788µs	6.6%
ANALOG INPUTS	2 ms	48µs	288µs	2.4%
			2864µs	23.9%

In order to present and discuss the microprocessor performance in this typical remote configuration a series of bar graph type illustrations has been compiled to represent the internal activities of the microprocessor. The vertical bars in Figure 3 indicate that the corresponding firmware module is using the processor for a period of time that is represented by the width of the bar. The time base for this series of bar graphs is chosen to be the scan frequency for status data acquisition (12 milliseconds) since if the processor has idle time within its fastest scan cycle requirement then it has time to perform all required functions. A bar graph is provided for each of the four critical functions presented above as if it were the only active function.

By combining the four bar graphs of Figure 3 into a composite bar graph with proper consideration for the priority of the functions, the bar graph shown in Figure 4 results. The first three milliseconds of the previously defined 12 millisecond interval are shown in the composite bar graph to more readily demonstrate the relationship of the functions to processor time.

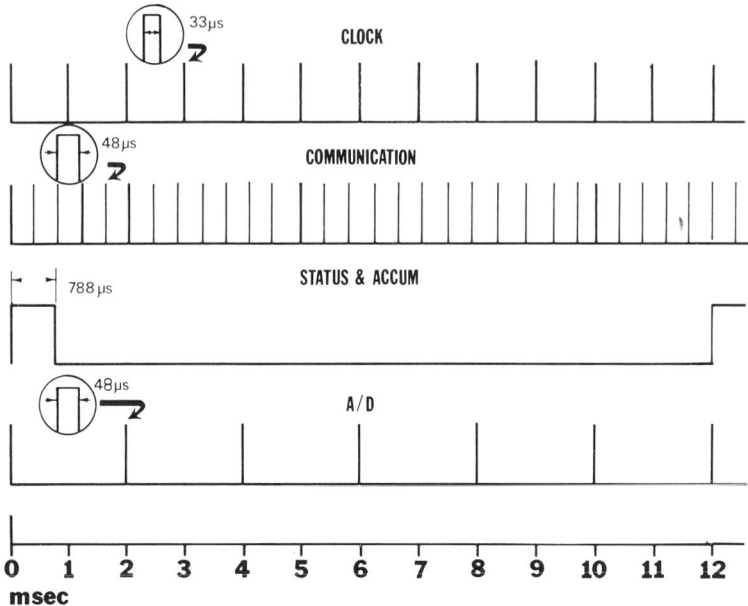

Fig. 3   Basic Timing Elements

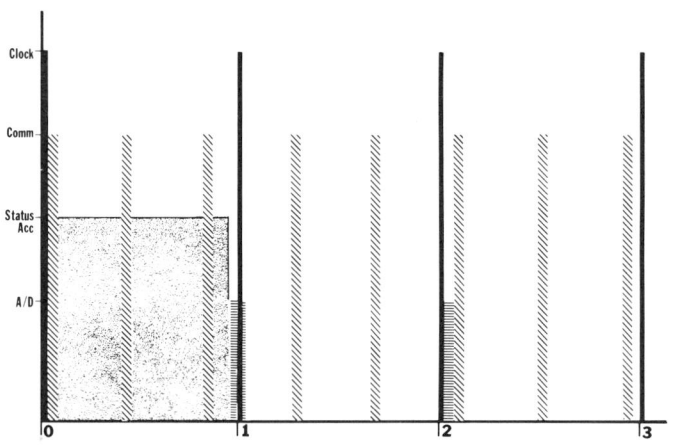

Fig. 4   Composite Timing

It may be noted that, due to the priority of the functions shown, the elapsed time required to perform each function is greater than the actual execution of the individual function due to interrupting events. For example, the stand alone execution time of 788 microseconds required for status processing is expanded to 965 microseconds of elapsed time because of the interrupting events of clock and communication servicing. Thus, the loading of a processor in an intelligent remote must be evaluated with such considerations as function priorities, interrupt driven functions, peak activity scenarios, and peak loading.

A final look at Figures 3 and 4 reveal that this high performance microprocessor has approximately 75% idle time during eleven of the twelve milliseconds in the analysis interval. If the ratio of 788 to 965 microseconds is used as a measure of the overhead (approximately 25%) an interruptable task will carry, then this processor can provide 50% of its compute power to additional tasks without compromising the mandatory SCADA functions of communications and data acquisition.

An intelligent remote may be designed around a medium performance microprocessor or around a high performance microprocessor. The former approach may require more functions to be performed by the hardware such as status change recognition rather than firmware whereas the latter approach can support simplified hardware design. In addition, the higher performance processor provides the ability to handle larger configurations and provides the flexibility to add additional functions to the remote such as those discussed in Section V.

In selection of an intelligent remote unit, the potential user must carefully evaluate the unit in the same manner as he would a hardwired unit. In addition, he must pursue in depth the expansion capabilities in terms of point expansion, function additions, processor loading, and maintenance ease.

VII. FUNCTIONAL EXPANDABILITY

Expandability is defined in terms of processor time, memory, and I/O resources in order to support the addition of new points or functions to the intelligent remote. Point expansion (e.g., adding status, analog or control points) must be accomplished without any modifications to the control logic of the remote; i.e., the remote must possess an initialization feature which will enable it to recognize the new points, ideally without prompting from the master station and without software/firmware inputs at the remote. A self-initialization

routine should be capable of recognizing I/O points and
allocate data memory accordingly.

The basic functions of a fully automatic initialization
routine is described as follows. During the initialization
each I/O card address is read. The remote configuration is
determined by recording the I/O cards that respond and
establishing the data base structure. Then all I/O cards are
scanned and the data (status, analog, and pulse count inputs)
are placed in the data base. The final action is to set the
communication interface in the receive mode and initiate the
internal periodic scans of data via the analog and status
scan routines. When all initialization is complete, the
interrupt system (real-time clock and communications I/O) is
enabled and control is transferred to the executive.

After determining the resources of processor time and
memory required to provide the mandatory SCADA functions for
the maximum planned point count, the spare resources available for new functions may then be determined as described in
Section VI. Care must be exercised in the definition and
implementation of each new function to avoid compromise to
the basic functions, particularly during communications processing and at restart.

VIII. CONCLUSION

Specific performance capabilities and limitations of
intelligent remote terminal units have been defined and discussed. High performance 16 bit microprocessors with memory
cycle times of less than 200 nanoseconds are available for
application as the control logic in intelligent remote terminal units. Speed of the processor is important, but only
in a hardware environment that provides a proper functional
allocation between hardware and firmware logic. The proper
allocation must take into consideration the duty cycle
associated with the function under analysis. The hardware
interfaces for standard functions like communications I/O
should be redesigned to reduce the number of interrupts per
second when additional functions such as Fault Data Capture
and Sequence of Events Reporting are to be included in the
intelligent remote terminal unit.

The user must decide the relative priorities that are
to be assigned to his functions. Some of the mandatory SCADA
functions (status and analog data acquisition) may be assigned a priority lower than Fault Data Capture or Sequence
of Events Data Capture. The duration of a disturbance may
therefore result in suspension of normal SCADA functions for
a few scans. The acquisition of the fault data analog para-

meters of current and voltage is temporarily given priority over the acquisition of the steady state watts and VARS. Only the user of an intelligent remote can properly establish the relative priorities of the functions that are to be included in that unit. This chapter has provided a framework and illustrated the analytical steps required to properly define the hardware interfaces and to assign appropriate processor priorities to each function that is to be included in the intelligent remote terminal unit.

IX. REFERENCES

(1). IEEE Working Group 72.2, "Manual, Automatic, and Supervisory Station Control and Data Acquisition Equipments," IEEE Standard (Pending), February 1, 1977, pp. 5C5/D2.
(2). Koenig, D.F., et al TRW and SCE, "Instrumentation and Control in EHV Substations," IEEE T 74 492-5.
(3). Laycock, G.B., TRW Controls, "Inside A Programmable Remote Terminal Unit," IEEE Region 5 Control of Power Systems Conference Paper, March 14-16, 1977.
(4). Street, A.G. and Gaul, J.A., Philadelphia Electric Co., "Microprocessor Programmable Remote Terminal Unit for A SCADA System," PICA 1977.
(5). Phadke, A.G., et al American Electric Power Service Corp., "A Digital Computer for EHV Substations: Analysis and Field Tests," IEEE F 75 543-9.
(6). Koenig, D.F., et al TRW and SCE, "Computerized System Fault Analysis," IEEE C 73 349-0.
(7). CIGRE Study Committee No. 34 Colloquium in Philadelphia, "Special Report For Subject 1 Trends In Automation In HV and EHV Substations," October 1975.
(8). Boeing, R.Milligan and Nilsson, Stig of EPRI, "Digital Fault Data Acquisition and Recording In Substations," COPS Paper.

IMPLEMENTATION OF THE SCADA CONTROL
CENTER FUNCTIONS

Max D. Anderson

University of Missouri - Rolla

I. INTRODUCTION

Earlier work on energy control center systems has been summarized in an excellent report by T. E. DyLiacco[1]. In this reference, the state-of-the-art of control center design is reviewed. Basic information is presented on control centers currently in operation or nearing operational status. A tutorial overview of control centers was subsequently prepared by DyLiacco[2], which describes the objectives, functions and elements of control centers. Emphasis was placed on the monitoring and control functions related to the security of the power system operation.

This paper presents an overview of the requirements and the rationale for the design and implementation of the following functions which make up a Supervisory Control and Data Acquisition (SCADA) system:

1. Data Acquisition
2. Data Base
3. Man-Machine Interface
4. Operating System
5. Performance Monitoring - Error Detection, Failover, Recovery, and Automatic Restart

The objective is to provide overview coverage of the SCADA control center functions for continuity and perspective and to concentrate in more detail on new innovations and problem areas associated with the design and implementation of a SCADA system.

A baseline hardware system is needed for reference in the discussion. A typical one is shown in Figure 1, where it should be noted that "dual" or redundant hardware is provided. To achieve the necessary availability of 99.5% or better for the SCADA system, it is generally required that every primary hardware subsystem or module have an identical hardware unit for "backup". In case of failure of the primary unit, failover to the backup unit and recovery of operation is achieved. How this can be implemented is the subject of Section VI.

The SCADA system must perform effectively and reliably as an operational system in an operating (Dispatching) environment. To obtain this goal, a well engineered and integrated hardware/software system is required. A logical approach to SCADA implementation is to define and develop the functions to be performed. These were listed above. Figure 2 depicts the inter-relationship of these functions with due consideration for the state-of-the-art hardware and software constraints. SCADA is comprised of functions (1) through (5), with few or

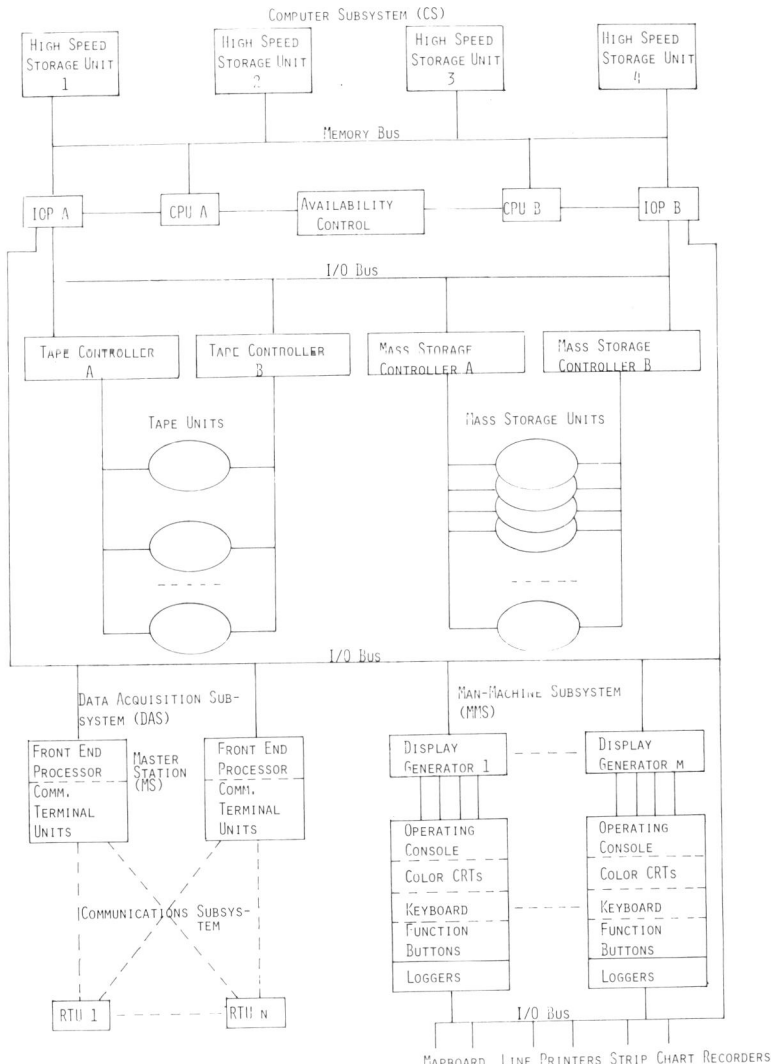

Fig. 1 SCADA System Hardware Diagram

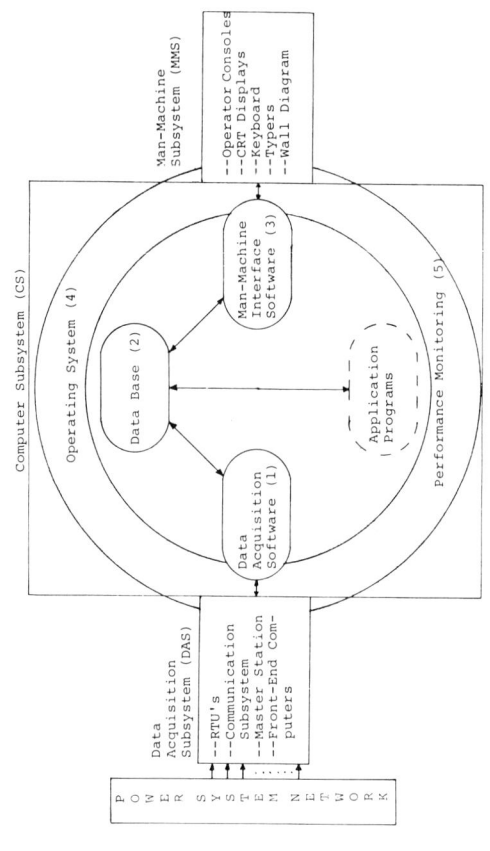

Fig. 2   SCADA Functional Diagram

no Application Programs (shown in the dashed oval). These functions are developed in the following sections of this paper.

The addition of Application Programs enables Energy Management System (EMS) or Energy Control Center System (ECS) functions to be performed. Since most SCADA systems eventually grow into EMS or become part of an EMS, it is worthwhile to discuss the SCADA implementation in the same context used in EMS to take advantage of the commonalty that exists in implementation.

II. DATA ACQUISITION AND CONTROL

The Data Acquisition function is required to:

1. Read the power system measurement data from 100-250 Remote Terminal Units (RTU's) into the control center computer under program control. Each RTU, located at a remote station, averages 20-50 analog points, 50-75 status points which are ready every 2 seconds, and 5-10 accumulator points read every hour.
2. Detect and handle data error conditions due to RTU and communication system malfunctions and noise.
3. Perform limit-checks on data based on high/low operating limits, thermal limits, safe loading limits, etc.
4. Scale and convert analog data in binary form to engineering units in floating point format directly usable by the other computer programs.
5. Interface with the Data Base Manager (DBM) which generates data base addresses, and store data in data base.
6. Store only good (error free) data in the data base. Bad or erroneous data should not be stored in the data base as they will cause failure of the user programs. Instead, quality indicator flags should be set to denote error conditions, retention of the previous value or use of estimated value.
7. Complete in time less than 2 seconds the above tasks for each scan of the measured data so that the next scan (2 seconds later) can begin.

Software design and implementation to satisfy the above requirements is described in the following paragraphs.

Data Acquisition is implemented by Master Station (MS) software and the Computer Subsystem (CS) Software. Two or more minicomputers (1 or more on line with one backup) are used in the MS for "front-end" processing to relieve the load on the CS; the latter includes 2 medium-large size multicomputers or multi-processors. (See Figure 1). Through the MS software, the minicomputers: 1) control the scan, 2) read the data from the RTU's, 3) check the data for errors and reasonableness against operating limits and set appropriate flags for the error handling and failover functions, 4) scale and convert the analog measurements into engineering units, floating point format, which is directly usable by system and user programs. Upon completion of the above, the data resides in MS (minicomputer) core, in the order in which it was scanned.

Next, the CS software through direct memory access into minicomputer memory, reads the data and maps it into the data base, using precomputed data base addresses. In addition, the CS software checks the error flags and provides the necessary interface to Alarm Generation and Performance Monitoring software for the handling of any errors or error conditions.

The MS software generally consists of two primary routines, 1) the Interrupt Handler (INTH), and 2) the Executive/Scheduler (EXEC). The INTH allows the minicomputer to operate in slave mode to the CS, which writes-with-interrupt its commands into the minicomputer's Command Table. INTH also initiates tasks in response to interrupts from external devices including timers. Upon completion of every command from the CS, the MS sends to the CS a completion response word which identifies any errors encountered. Through the analysis of these error conditions and timing out the response word, the CS can assess the "health" of the minicomputers, and initiate failover when a malfunction has occurred.

The EXEC program provides for completion of tasks assigned to the MS such as the scanning/polling of the RTU's and the servicing of peripheral devices attached to the MS. The EXEC operates in a continuous loop checking flags in the Executive/Scheduler Tables to determine what processing remains to be done. Many of these flags are set at the interrupt level, so that the logical processing flows from the interrupt handling (INTH) routines to the executive (EXEC) routines and back again.

The scan/polling of the RTU's is timer initiated in the CS every 2 seconds by a "Prepare for Scan" command sent to each minicomputer. If the minicomputer has completed its previous scan, and there are no outstanding commands from the CS, then each Nova returns a "Ready for Scan" response to the CS. In response to the "Start Scan" command from the CS, the Nova issues a "snapshot" command to all the RTU's. This is done in such a manner that each RTU will read all of its analog, discrete, status, and accumulator measurement points within 10 msec., convert the analog measurements into unscaled binary, and store the values in the RTU's memory.

The EXEC can now initiate the reading of data from the

first RTU on each of the communications loops. One RTU at a time, all the data is transmitted serially through the Communications subsystem and into the Novas at the rate of 2400 bits/sec. Because of geographically separate communication paths (loops) to each station (RTU), the MS receives duplicate data from each RTU, which are then compared and error checked. The appropriate error information is entered into a Path Status Table, which will become part of the Data Base.

A single copy of "good" data is checked in the MS for status changes and high/low limit violation, scaled, format converted, and stored by station in the following order in the Static Table:

RTU #i   Loop #j   for i = 1, 2, ..., no. of RTUs
                       j = 1, 2, ..., no. of comm. paths

1. Changed and current status bits
2. Discrete Data
3. Analog Data
4. Accumulator Data

This format is repeated in sequence for each station (RTU) on each communication loop.

When the EXEC has determined that all data from the RTU's has been read into the MS, it issues a scan completion response to the CS. The CS will then read the Static Table and the Path Status Table from the minicomputers via direct memory access, resolve the addresses, and store the data into the Data Base.

Support software includes a Cross-Assembler to enable the MS (minicomputer) operational software to be assembled in the CS. Multiple minicomputer core loads can be assembled and stored in the CS as changes and additions to the MS software are required. An on-line upgrading capability is generally provided; i.e., after a new minicomputer program has been assembled and debugged, it can be loaded into the Nova by the CS and begin execution on the next scan cycle.

Tradeoffs have been made for implementing the data acquisition functions in a separate front-end computer system versus implementation in the main CS. Generally, it has been determined that:

1. Front-end minicomputers can execute the relatively short programs for the scanning and preprocessing of data, which are repeated every 2 seconds, more efficiently than can a large scale computer of the type used in the CS.

2. The overhead of operating systems presently available for large scale computers such as those used in the CS, makes the data acquisition very inefficient in the CS, and greatly increases the execution time of other system and user programs of lower priority. This is due to the fragmentation of the time and memory allotted to the larger application programs caused by the data acquisition routines.

3. Front-end minicomputers add a higher degree of availability to the system through their redundancy especially when critical operating displays can be driven by the front-end computers directly with scanned data, in case of a CS failure.

In the near future we can expect to see the use of microcomputers in the MS front-end processing, where they can best perform some of the aforementioned Data Acquisition functions. We can also expect to see microcomputers used in RTUs to do some preprocessing of the measurement data and reduce quantity of data transmitted over the communication subsystem from the RTUs to the Control Center computer system. Reference 3 discusses some of the functions that could be performed in an Intelligent RTU (an RTU with microcomputer) such as a) event sequence recoding, b) near continuous monitoring of analog points and transmitting values when changes occur, c) measurement data preprocessing. Performance benefits for the Data Acquisition functions are also described along with a functional design of an Intelligent RTU.

III. DATA BASE

The effectiveness value of a SCADA system is primarily a function of the data processing efficiency of the digital Computer Subsystem (CS) as measured by data thruput. Operating in a real-time environment, the CS controls and processes data which flows between the power system network/equipment and EMS operator consoles and printers, Figure 1.

To enhance the data thruput, the SCADA system requires an efficient and effective Data Base subsystem. From the data thruput point of view, the Data Base is the heart of the system. It is the data interface for all functions; i.e., application programs, data acquisition, and man-machine functions, Figure 2. The primary responsibility of the Data Base in this role is to provide a well ordered and well defined process for these inter-program data transfers. Hence, it constrains the implementation of nearly all functions, and thus affects the utilization and performance of the computers' processing power.

Specifically, the Data Base is the repository for all

power system dynamic data, common data used by more than one application/user program, manually entered constants and variables from the operator consoles, and network data including historical files and records of these data. It physically resides within the CS and its peripheral devices. Generally two storage media are used:

1. high speed storage (core memory and/or IC register) for dynamic data including data with a high frequency of access (near data acquisition system scan rates).
2. file oriented mass storage (disk or drum) for network data, variables, constants, and historical data, i.e. data items with a low frequency of access.

A survey of the present state-of-the-art in data base development is presented in Reference 4, which includes an extensive list of references and an explanation of their content. Reference 4 is general to a large class of data bases. In contrast, Reference 5 attempts to focus more closely on the real-time operational aspects of data bases for EMS. This section attempts to summarize the findings in Reference 5 for SCADA systems.

A. Data Base Requirements and Design Objectives

A summary list of data base requirements is presented below. These also may be considered design objectives for the data base:

1. Fast access during program execution
   a. single level addressing
2. Minimum overhead during execution
   a. addresses resolved off-line (not during mainstream execution)
3. Facilities for application/user programs
   a. data accessing buffering and ordering
   b. program independence from DB
4. Data base manager facilities for
   a. accessing data (pointer tables)
   b. data base security
   c. maintenance and upgrading (interactive).

Data transfers within the computer system must be fast in order for core resident computer program processing to yield results that are meaningful in a real-time sense; i.e. up-to-date. Since data transfers directly involve the data base, a major portion of processing overhead within the computer will be devoted to accessing (reading or writing) individual data points within the data base. Any reduction in data point accessing time will reduce data base processing overhead and hence contribute directly to enhancing real-time operation.

The requirements for capability to easily change or restructure the data base look to the future demands placed upon the SCADA and computer subsystem. Power system network configurations are not static. Transmission lines, substations, generating stations, etc. are added to or deleted from the power system. As equipment is upgraded, limits, network characteristics and data points are changed. These changes will generate additions or deletions in measurement data, network data, constants, variables, etc. in the data base. Also contributing to these changes are the changes to, additions of, and deletions of application programs. Therefore the overall structure of the data base is dynamic. It must be flexible and easy to change.

The objective of meeting these requirements simultaneously in a Data Base is not without its problems. The requirements of fast data access and ease of restructuring work at cross purposes with each other. If fast access is achieved, the result tends to be an inflexible Data Base structure. If a flexible structure is achieved, it results in slow access. Thus compromises which will generate data base characteristics approaching both goals must be made.

To formulate an approach for obtaining a solution to these problems, which will satisfy the above requirements, we must investigate the ingredients of data base design.

B. <u>Ingredients of Data Base Design</u>

The ingredients of data base implementation can be classified into four task areas:

1. content and naming convention of the data base,
2. data base accessing methods,
3. structure of the data base, and
4. the data base manager software.

1. Content and Naming Convention

The first point of consideration is the contnet of the data base. A definition of all data item names and descriptions of every data item needed in the EMS must be made. This is a formidable task which should be assigned to a specific

SCADA CONTROL CENTER FUNCTIONS

group of persons within the utility company representing all departments involved. It is the responsibility of the Operations and Applications groups to determine, on a data item basis, what data is needed for power system operation and for data input to the programs. They should also define the required data characteristics of accuracy, resolution, sampling period, range, format, etc. It is the responsibility of the Engineering and Station Design groups to supply the data via RTU and communications equipment to the required accuracy, resolution, sampling period, etc. In cooperation, these groups must put together an adequate description of all the data to be used in the system.

For power system data, the quantative definition of data can best be done on a station basis; i.e., for each generating, transformer, and switching station, all of the data items for the following data types should be identified:

a. Voltages
b. Line and station MW and MVars
c. Transformer tap positions
d. Breaker positions
e. Disconnects and grounding switch positions
f. Capacitor banks
g. Security status
h. Alarm status
i. etc.

For each data item in the above list of data types, the following characteristics should be defined:

a. Data item name
b. Description of data quantity
c. Scan period
d. Accuracy - transducer and ADC
e. Resolution/Precision
f. Error rate statistics
g. Range of values
h. Format
i. Scaling and conversion factors
j. Alternate source of data
k. Document and drawing reference.

This information forms the basis for interface control documentation.

Related to the content of the data base is the naming convention of its individual elements. Data elements must be accessible by symbolic names that are applicable to all

language levels used. Starting with the power system data,
each data item needs a name sufficient to identify it by attributes, allow easy identification by the operating personnel and
programmers, and easy resolution of addresses in user programs
for rapid access of the data.

The naming convention should be based on existing naming
conventions used by operating personnel. Though this may not
be the most efficient approach from a data management point of
view, the data items names must be familiar to the operators,
because the names will have to be displayed in that form on the
CRT pictures to be useful to the operators, and data will be
manually entered via the consoles by data item name. Abbrevi-
ated names or mnemonics should also be defined for use by the
programmers, due to the constraints of the computer address
field used in their source programs. A good naming convention
can greatly simplify data base maintenance and the use of the
data base by application and system programmers. The following data attributes should be included; e.g., for power system
data:

    a.   Station name and kV level
    b.   Equipment name or I.D.
          - Transmission line (name or mnemonic)
          - Transformer
          - Capacitor bank
          - Breaker
          - Disconnect and grounding switches
          - etc.
    c.   Operating quantity (by data type or units)
          - MW
          - MVar
          - KV
          - MWH
          - Frequency
          - Status
    d.   Data characteristics
          - Measured via data acquisition
          - Computed or filtered by application program
          - Manually entered by system operator
          - Manual over-ride by system operator
          - Data quality and age indicators.

It is important to note that all data item names must be
unique.

2.   Data Base Accessing Methods

    The simplest and fastest method of addressing a specific

memory location is by single level addressing sometimes called direct addressing. Single level addressing implies a single read operation of memory to obtain the next operand. It is performed by calculating a relative address from the sum of the contents of a base register, possibly an index register, and a known displacement contained in the instruction address field.

A second method of addressing is indirect addressing. This method may involve multiple memory read operations within the procedure of calculating a final effective address. Hence, it is a slower memory access method. Therefore, enhancement of real-time EMS operation suggests that single level addressing of data base elements be used. The possible accessing methods are dependent on the Data Base structure discussed below.

3. Structure of the Data Base

The third point of consideration in data base design is the skeletal layout of the data base structure. Consider three approaches that can be taken and tradeoff their value versus computer time and high-speed memory costs:

    a. Data base structured for rapid access by user programs.
    b. Data ordered by station in correspondence with the scanning of data.
    c. Pointer table approach for data base addresses.

Consider the first approach (a). Ideally, the user programs should have the facilities to execute under single level (direct) addressing; i.e., to access a data base operand in a single memory read cycle. This minimizes the user program's access time to input data contained in the data base. An example layout for this type of data base is the three dimensional array shown in Figure 3. Any data item can be identified in the source language by the triple of numbers (category, article, type) and actual addresses resolved in executable code by the compiler and mapping (link-edit) processors.

This approach enables data items to be accessed in minimum time individually or by groups by using a comprehensive set of data attributes (and/or by symbolic name) consistent with the naming convention. However, accommodating the user programs with rapid access to the data base adds computer time and memory expense to the EMS because of constraints on the Data Acquisition subsystem and the CS overhead as explained below.

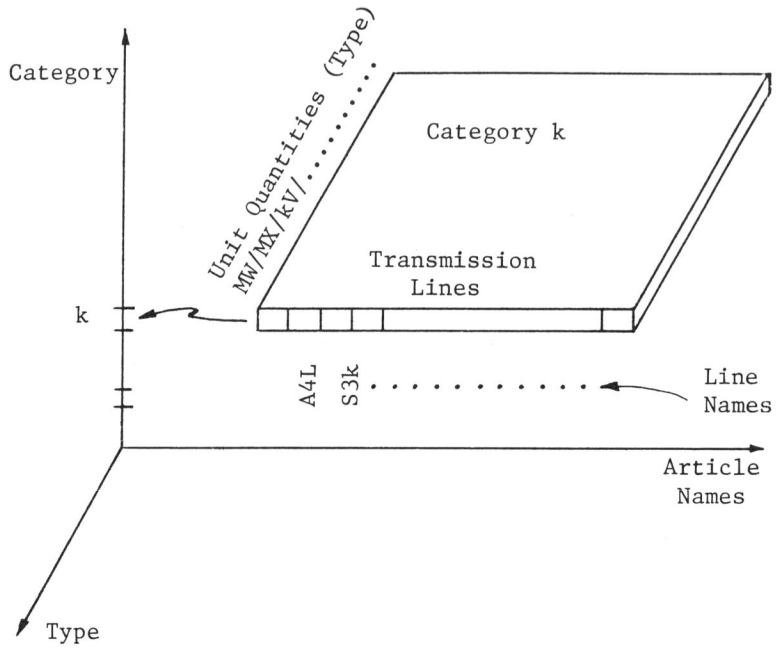

Fig. 3   Data Base Layout for Minimum Access Time

This gives motivation for the second approach (b). State-of-the-art remote terminal units (RTU), modems, and data communications facilities provide for data to be input into the computer system in the order in which it is scanned or polled from the RTU's; namely, from one station at a time and in the order stored in RTU memory. All the data from one station is read in serially, then the same sequence is repeated for the next station and so on, until all the data from the stations have been read into the EMS computers, completing the scan sequence. Hence, the data is normally ordered by station and not by data type, and it is processed by the data acquisition function in this order.

To reorder the data by type or put it in any other order or grouping for the user programs requires another mapping in the CS, i.e., another pass through of reading and writing the data from one area of memory to another. This requires more computer time and memory.

The trade-off consideration is that not all user programs require all the data from every scan. For example, assume the data is scanned every 2 seconds. State estimation may execute only every 1-15 minutes. CRT pictures may be updated only as data changes occur. Therefore, from the computer utilization point of view, it may be less costly in time and memory to have the user programs reorder the data in the form required for processing than to have it reordered for them by data acquisition or the data base software.

The third approach, (c) is a compromise which allows an improvement for the user who is constrained to go with the second approach (b) data ordered by station. It is to define a table (or tables) of pointers to the data base, entered by data item name, which points to the data base address of the data item as shown in Figure 4. This approach is not as efficient as the single-level addressing approach (a). As shown,

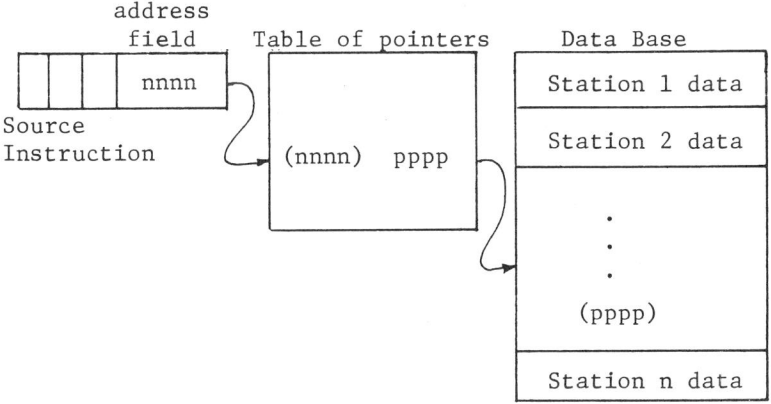

Fig. 4 Third Data Base Design

the access time to a single data item is at least doubled. Depending on the naming convention used and the size of the data base, it may be desirable to have multiple levels of pointer tables to select sets of data according to certain data characteristics or attributes and to accommodate matrix computations.

Three additional factors favor the third approach (c) of using pointer tables. First, the pointer tables make the user programs independent of the data base in that the only addressing done by the programmers is by data element name according to the naming convention. Secondly, the data base can be made secure by controlling access to it through the pointer tables; i.e., password check of the calling program to prevent unauthorized programs from corrupting the data base. Third, the building of the pointer tables can be automated by a data compiler[6] type program, (which is part of the Data Base Manager software) such that the user need not be affected when the data base is expanded or reordered.

Regardless of which of the structures is used, part of the data base will reside in fast access memory and the remainder in mass storage. A decision must be made on how the data points will be apportioned to these storage media. This decision is directly influenced by the speed of access of selected data points since retrieval of data from mass storage is orders of magnitude slower than data retrieval from core. For example, IBM's virtual storage systems seems to have an advantage in that data accessed frequently is automatically moved from mass storage to fast access storage and retained there until the frequency of access decreases.

4. Data Base Manager/Maintenance Software

The fourth point of consideration in data base design is the creation of a Data Base Manager. The Data Base Manager (DBM) is a software package which is part of the Executive software and may be integral to the operating system. It exists for the purpose of controlling and directing all software and hardware processing associated with the Data Base (DB). The operation of the DBM can be categorized into four general functions:

1. maintenance of directories of all data points stored in the data base. These directories contain the attributes of each data point necessary for efficient data location and processing. These attributes include the elements' symbolic names, memory addresses, storage format, etc.
2. control of the physical layout of the data base in memory. This function includes determining where (core or disk) a data point is stored and its relative address. It is implemented under control of the operating system.
3. add or delete data elements from the data base. Additions or deletions of data points occur on a dynamic basis as the power grid topography and components are changed. Interactive capability within the DBM should be provided to enable this function to be implemented by electric utility company operating personnel.
4. interface all computer operations with the data base. The primary concern of this function is to maintain a consistent data set within the DB and provide programs requesting data with consistent data subsets. This function includes providing the necessary read/write locks for security of the

data base.

These categories comprise the four functions of the DBM. The following paragraphs will discuss the implementation of these functions.

The first two categories, maintenance of the directories and control of data base layout, are directly concerned with creation of the data base. Let the first data base directory be called the Symbolic Name Element Directory (SNED). It is, to a large extent, created by the user personnel as an early task in implementing the EMS. That is, the utility company creates a list of all data points needed by EMS operations, assigns symbolic names to the data points, according to the naming convention, and partially builds the SNED. The building of the SNED consists of supplying information about the data element attributes. The number of data attributes should be kept to a minimum to enable the DBM to operate as fast as possible. Some of the data element attributes are supplied by the DBM during its creation of the data base within the EMS computer. The data element attributes are listed below:

1. symbolic name
2. storage requirements; the number of memory bytes or words devoted to the element
3. storage media; core or disk
4. write locked data group membership number; this attribute identifies which data group the data element belongs to for purposes of write locking the element (write locking is explained later)
5. read locked data group membership number; this attribute identifies which data group the data element belongs to for purposes of read locking the element (read locking is explained later)
6. current relative address on its storage media. Attributes (1)-(5) are supplied by the power company personnel. Attribute (6) is supplied by the DBM during the creation of the data base.

The creation of the DB within the computer is preceded by booting of the computer operating system followed by the activation of the DBM. The initial input of the SNED to the DBM will most likely be by a card deck. Appropriate JCL and data card formats will conform with operating system and DBM requirements. The DBM will store this partial table on disk, and thereafter, during system re-boots, it will be read by the DBM for processing in lieu of re-reading the card deck.

The physical layout of the data in memory will be a

function of the particular storage structure (as discussed previously) chosen by the creators of the DBM. Hence, the program logic of DBM will be created to reflect this choice. By using the data attributes (2) and (3) of the SNED, the DBM will make known to the operating system, its storage requirements for the data base. The operating system will calculate and identify to the DBM appropriate base register contents and relative displacements of the membory location addresses allocated to the data base. This information will be appropriately stored by the DBM in the SNED attribute (6). Initialization of the memory locations will be through appropriate software and hardware of the EMS system and application routines. At this point, the DB has been created, and is ready to begin its functional performance in the EMS.

A second directly can be used to contain additional data attributes. These attributes are characteristics of data elements not directly involved in addressing element values as are the attributes of the SNED. They will be used primarily for enhancing input/output and display of data element names and values on operator console CRT's, line printers, mapboards, etc. Specifically the data attributes maintained in the second directory are: data element names (defined in agreement with the naming convention), point description attributes, program I.D., file I.D., picture I.D., as applicable.

The third function category of the DBM, addition and deletion of data elements, is mainly a replay of the creation of the data base. This replay may require a major or minor processing depending on the current data base layout in memory. Again, the processing of the addition or deletion of the data points is handled by the DBM in cooperation with the operating system; however, it can be performed and tested in the back-up computer or in background made so as not to interfere with the real-time processing. Capability should be provided for the data changes to be made conveniently through a CRT console keyboard as well as by card input. That is, interactive capability with the DBM should be provided through the operator consoles and the data entry software. Hence, the DBM will have to be able to recognize and handle these different data entry formats.

The DBM will request to the operating system reallocation of memory for the data base. This procedure may not require a complete remap of the memory by the DBM. The addition or deletion of a few data elements will not necessarily change the relative addresses of all data base elements and hence recalculation of these addresses should be avoided in order to reduce processing overhead. Therefore, the logic of the DBM should be carefully constructed so that it has the ability to examine the data base structure and determine the best procedure to request, from the operating system, new memory allocation in order to keep processing overhead to a minimum. After this operation, the SNED will be adjusted to reflect the new and/or old element attribute states.

The fourth category of DBM functions involves the general input/output procedures of transferring data from the data base to the user programs and from the user programs into the data base. These procedures must satisfy two objectives; which are potential problem areas:

1. to avoid recompiling and remapping the user programs every time a change is made to the data base, the user programs must be made independent of the data base layout and addressing scheme. The application programmer should have to be concerned only with data element symbolic names, and he should have the flexibility to order his input data (from the data base) to minimize processing time and overhead.
2. The data transferred must be a consistent data subset; i.e. data taken from the same sample time.

The input/output procedure is shown in simplified form in Figure 5. Each program that requires input data from and/or output data to the data base is assigned an individual input

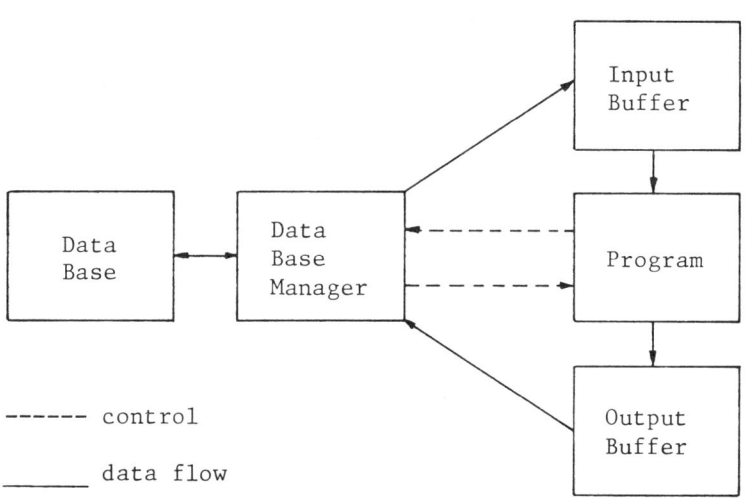

Fig. 5   Data Base Input/Output Procedure

and/or output buffer in memory that serves as a repository for that data. During the input of data to a program from the data base, the program requests all of its required data at the same time. Control is passed to the DBM in a fashion similar to a subroutine call, and the DBM collects the required data values from the data base and places them in the program input buffer. Then control is returned to the program. As the program executes, it uses data values from its input buffer as needed. Any data values needed to be stored by the program in the data base are deposited in the program output buffer. When the program has filled the output buffer with the required data values, it again passes control to DBM. The DBM then controls the actual transfer of the data from the output buffer into the data base.

IV. MAN-MACHINE INTERFACE

A. Requirements

The Man-Machine function is required to provide the following capabilities and facilities. These requirements are grouped in correspondence with their software implementation described in Section B:

 1. CRT Displays - Display pictures with power system information on the operator console color CRT's. The following CRT display picture types are required:
    a. Generating, transformer, switching-station one-line diagrams
    b. Transmission line diagrams
    c. Tabular data displays
    d. Trends - graphs of variables with time
    e. Barcharts
    f. Data Entry forms
    g. Menus - for selection of CRT pictures, data logs, and initiation of application programs.
 2. Data Entry - via console keyboard, function buttons, and CRT Data Entry pictures. Requirements include:
    a. Requests for CRT pictures
    b. Initiation of control actions - initiation of programs into execution (on demand) which can automatically request CRT pictures for operator interaction in performing an operating procedure.
    c. Manual data entry into the data base (Section III)
    d. Data Entry forms - for passing parameters to a program.

All data entry operations may be selectable by cursor from a menu of options or selection points on a CRT picture as well

SCADA CONTROL CENTER FUNCTIONS

as by function buttons and console keyboard entries.

3. Wall diagram, strip chart recorders and other special analog and digital displays.

These devices are listed separately, because they are output only and are physically separated from the operator consoles.

Their requirements are based on adopted operating procedures which vary between utilities. The principal requirement is that SCADA drive and update these devices efficiently and with minimum time delay.

4. Alarm Generation - An alarm condition occurs when the power system goes into an emergency state (see Reference 1) for which corrective action may be required. When an alarm condition occurs, it must be properly detected and identified. The operator must be notified with an audible alarm and a short message displayed on his alarm CRT. This message must contain sufficient information to enable the operator to properly identify the alarm condition, evaluate its threat to the security of the power system, and decide on the corrective action to be taken.

An important requirements definition task is to define the alarm conditions to be checked and determine where and how to perform alarm condition detection and identification. Since alarm condition information is distributed throughout the system, this task should be included in the requirements definition task for each SCADA function. A second part of this task is to determine how best to present alarm information to the operators to obtain their most effective response to the alarm condition.

5. Data Logging - Provides the management information system facilities for the power network. The following types of hard copy reports are required:
   a. Periodic reports to include hourly, daily, weekly summaries of power system information
   b. Reports of critical and summary information on demand
   c. Customized reports of critical variables as selected from the operators' consoles
   d. Logging of key/critical events in time sequence, including alarm messages and error messages
   e. Operator action logs which are time sequence loggings of operator actions at the consoles including responses to alarm conditions and manual data entries.

6. Picture Compiler - This facility becomes a requirement when the number of CRT pictures becomes large, for example > 400, or the operators need to be able to create and/or modify CRT pictures from the consoles. The two types of picture compilers suggested are

a. Card input - for entering pre-defined CRT pictures into the computer system
b. Interactive - for creating and/or modifying CRT pictures from the operators consoles.

Several general requirements apply to the above functions:

1. The data displayed on the CRT pictures must be current. Therefore, dynamic updating from the data base must be provided as new measurement data is obtained via the Data Acquisition function.
2. The response time of the computer to operator console requests for pictures should be in 1-3 seconds.

B. Implementation

The man-machine interface is implemented in the Computer Subsystem (CS) as a set of integrated programs and routines which fulfill the above requirements. These are described in the following paragraphs.

Note that the Supervisory Control functions are really implemented by the CRT Display functions and Data Entry. For example, the control/operation of a power circuit breaker (PCB) or motorized disconnect switch, is performed by the following procedure:

1. The PCB or switch is cursor selected from a station one-line diagram displayed on a CRT. The "select" command may be transmitted to the appropriate RTU to place that control point in the "selected" mode.
2. Through the Data Acquisition and CRT Update functions the operator is able to verify that the proper control point has been selected.
3. From the same CRT Display picture the "open" or "close" operation is commanded by cursor selection or console function button. The control command is transmitted to the appropriate RTU to operate the selected control point.
4. Through the Data Acquisition and CRT Update functions the operator is able to verify that the selected equipment has operated successfully.

1. CRT Display

The CRT Display function responds to a picture request from the Data Entry function (Data Entry) as shown in Figure 6. The CRT Display program retrieves the picture information

SCADA CONTROL CENTER FUNCTIONS

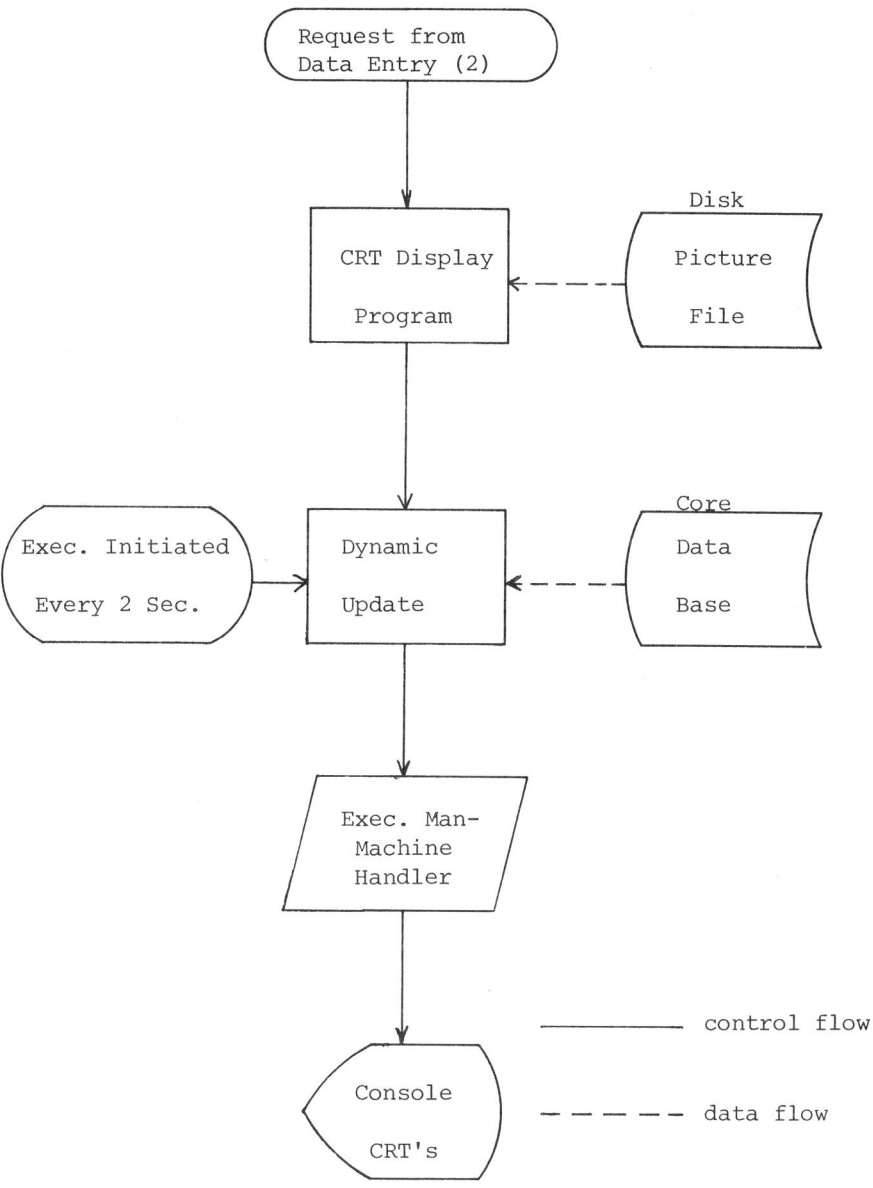

Fig. 6 CRT Display Function

from the Picture File (on disk). The Dynamic Update Program reads dynamic data (MW, MVar values, breaker status, etc.) from the data base. Both the static and dynamic portion of the picture are assembled as a character string and transmitted to the CRT via the Man-Machine Handler (V-B), which provides the necessary driver interface with the man-machine subsystem hardware.

The Picture File, which has been created by the Picture Compiler, is composed of variable length records, one record for each picture frame. Each picture frame record may contain up to 3 tables, which are:

    a. Control Table - Picture ID and control information including program linkages with picture frame.
    b. Dynamic Table - Contains linkages to the data base for retrieval of dynamic values and status to be displayed, and information for use by Data Entry in responding to operator entries on the picture frame being displayed.
    c. Static Table - Contains the static picture elements which may be transmitted to the man-machine hardware without further processing.

If a picture frame contains dynamic data, then the Dynamic Update program reads the data base and transmits new data values and status to the CRT Display, after each updating of the data base by the Data Acquisition cycle (2 seconds). This is facilitated by retaining in core the Dynamic Table for each picture frame being displayed. (About 4K bytes of core per CRT is required).

An application program may also request (but not enable) a picture to be displayed, for which it is providing the data, or through which it asks for data (via a Data Entry form display). To minimize interruptions to the operators, a hard and fast rule is that "A picture shall not be displayed unless requested by the operator". Conformance is provided for by a "deferred display" message to the operator identifying the program and/or type of display being requested by a program. Operator keyins enable the deferred displays to be shown on the CRT's. The exceptions are alarms and response messages to the operator, which are displayed on the Alarm CRT or the top 1-3 lines of any CRT as soon as they occur.

2. Data Entry

The Data Entry function responds to keyboard and function button inputs from the Operators' consoles as shown in Figure 7. The Exec. Man-Machine Handler (V-B) services the interrupts from the console hardware and places the console entries into the Data Entry input queue. The Data Entry program

SCADA CONTROL CENTER FUNCTIONS

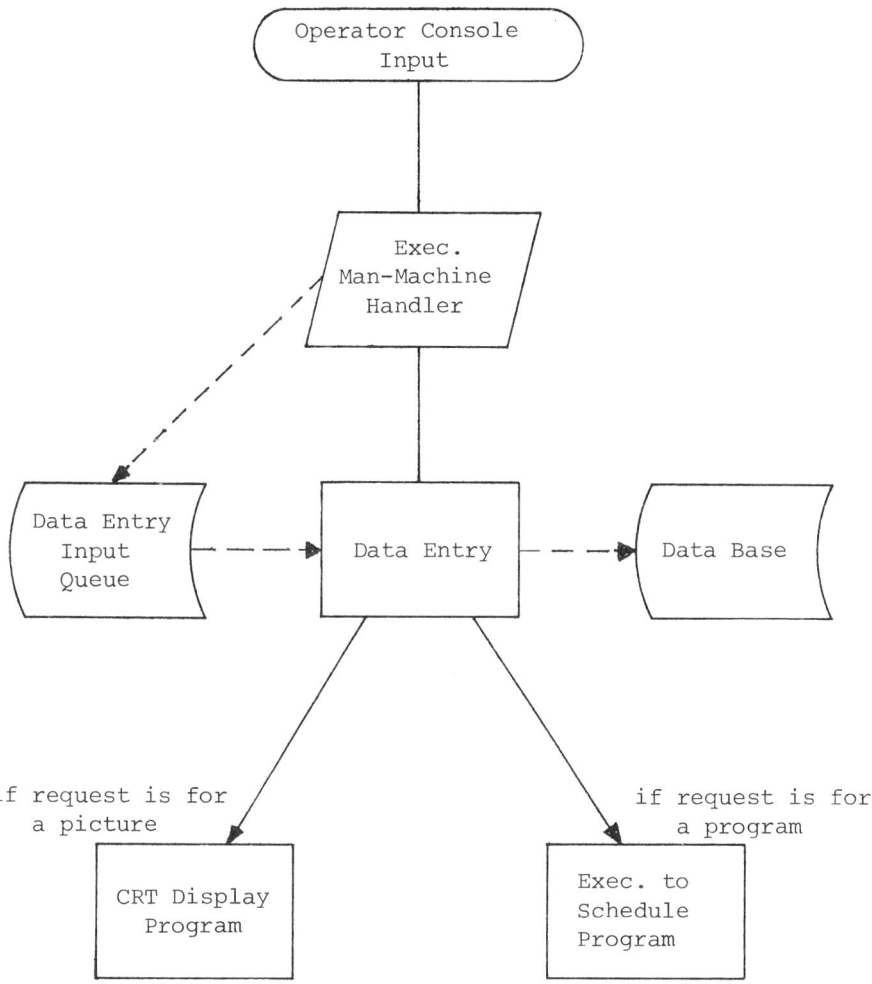

Fig. 7   Data Entry

processes these entries one at a time, passing the appropriate control and addressing information to the servicing program; depending on the Data Entry type:

    a. CRT Display program is scheduled and the appropriate picture ID and return information are passed to it, if the entry requests a picture.
    b. Exec. is called to schedule a program, if entry is a control action or requires a computation.
    c. For manual entry of data via a CRT picture (e.g. a Data Entry form), the Data Entry program makes use of

the Dynamic Table (in core) for addressing into the Data Base or passing the data entered to an application program.

3. Wall Diagram, Strip Chart Recorders and Special Analog and Digital Displays

These are output only display devices which must be updated each time the Data Base is updated with a new set of measurements. Software similar to the Dynamic Update program plus driver software is required for implementation.

In several systems, these devices were driven through the front-end computers to simplify the hardware interface.

4. Alarm Generation

Implementation of Alarm Generation consists of two distinct tasks: 1) the detection and identification of each alarm condition, and 2) the display of sufficient information, in the form of a short CRT message, to enable the operator to understand the alarm condition and take corrective action.

The detection and identification task (item 1) is implemented by software routines distributed throughout the SCADA system and application programs, where the data for the alarm condition initially exists. It is the responsibility of these programs to detect and identify alarm conditions and also to determine when the alarm condition has been properly corrected and no longer exists. The same implementation holds for error, response, or information messages to the operator. Upon occurrence of these conditions, the detecting program calls the Alarm Generation Program and passes it identifying information and data, if required, as shown in Figure 8.

The Alarm Generation program assembles a short and concise alarm message and initiates its display (item 2) to the operator on his alarm CRT. This is done in the following manner. First, it retrieves a skeleton alarm message from an Alarm Message File in the same manner the CRT Display program would retrieve a picture frame record from the Picture file. Alarm messages are normally (95%) one-line in length. Why? Because the operator doesn't have the time to read many lines of text to find out what problem he needs to solve. But provision is made in the software for multiple line messages, as some will be required.

SCADA CONTROL CENTER FUNCTIONS

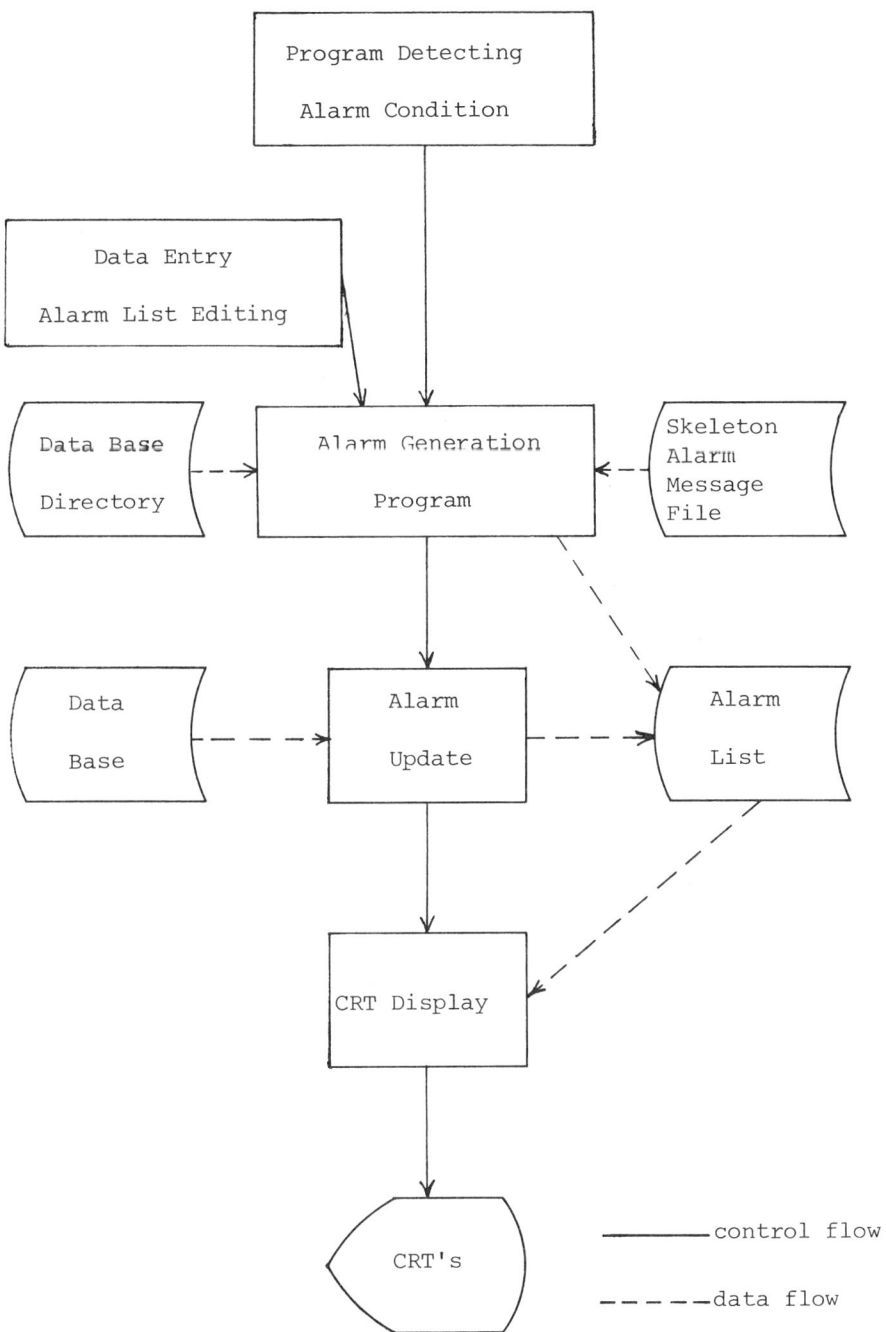

Fig. 8 Alarm Generation

An alarm message may contain the following type of information:

a. Time, date
b. Text message
c. Equipment names - from Data Base Directory
d. Dynamic data - values and status from data base
e. Limit values

The skeleton alarm message contains the text message, field and format definition for the variable information. As an alternate, this information can be passed by the calling program. Equipment names used in alarm messages must be in the standard naming system used by the operators.

For example, consider a line overload message:

Time	Text	Equip. Name	Text	Limit Value	Units	Text	DB Value	Units
HH:MM:SS	LINE	L4D	EXCEEDS LIMIT OF	340	MW	ACTUAL	354.2	MW

Alarm Generation program now assembles the complete message. Equipment item names are retrieved from the Data Base Directory (III-B). Dynamic data is read from the Data Base by the Alarm Update routine and inserted into the proper field in decimal format in the manner similar for CRT picture update. Alarm messages are also updated as Data Base values change. The time and date are added, and the completed message is added to the Alarm List, which is stored in fast memory. An audible alarm is also turned on.

CRT Display program is then called to display the Alarm list with the new message added. New messages are added at the bottom line of the Alarm CRT and the existing messages are pushed upward.

The Alarm List is the current list of existing alarm messages. It may exceed one CRT page; hence, provision for paging is provided. The operator has the capability of editing the Alarm List through the Data Entry function. 1) He may block alarm messages if he is aware of a problem which causes the recurring of similar alarm messages. 2) He may delete alarm messages for which the problems have been corrected and which have not yet been deleted by a time-out routine.

Alarm messages that have been deleted are removed from

the Alarm List and the CRT, and the space closed-up on the CRT screen.

Alarm messages are logged by the Data Logging program when they occur, and when they are cleared or deleted.

5. Data Logging

Implementation of Data Logging consists of: 1) The periodic collection of data from the Data Base to build a time history of data, and 2) the generation of logs and reports from the data collected. (See Figure 9).

Data needed for logs and reports is periodically collected from the Data Base via a Data Collection Program under Exec. Timer Control (V-B) and stored into the Data Collection File (on disk). This normally circular file is defined via a batch program which determines the data base addresses to be read and establishes start time, collection frequency, end time, or length of the data collection cycle.

A report generator program is used to define the reports to be generated, format the data, and produce print files for the designated printing devices.

Predefined reports are produced periodically (hourly, daily, etc.) under Exec. Timer control. They can also be produced on demand through the Data Entry program.

Customized reports can also be produced by defining the variables to be logged, frequency, start time, etc. through a CRT Data Entry form.

Alarm messages and Operator actions are also logged in time sequence. These can be printed on the console typers or the line printers.

6. Picture Compiler

The Picture Compiler may be implemented as 2 separate software modules: 1) Card input, and 2) Interactive picture compilers. The card input Picture Compiler (shown in Figure 10) is a language processor which provides for building the Picture File in the following manner:

   a. CRT Pictures are drawn on paper to scale in color on a standard x=72 by y=48 character grid for approval

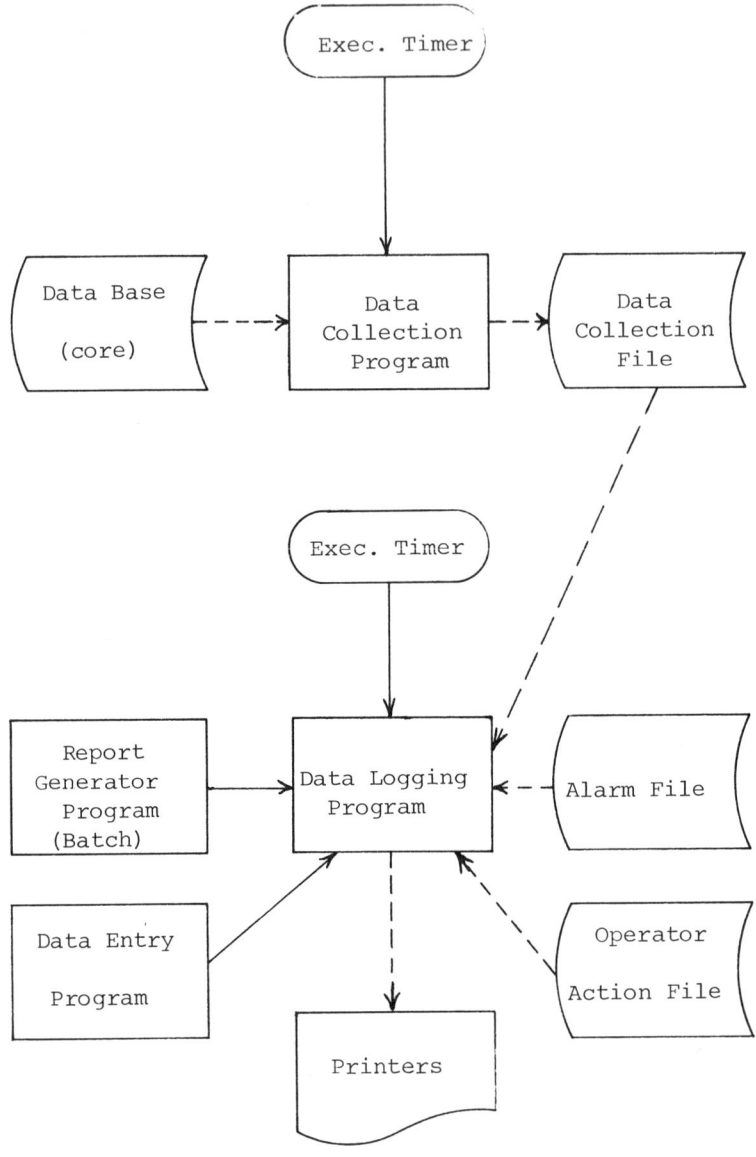

Fig. 9 Data Logging

    by the Operations Department.
- b. Pictures are coded in the picture compiler language, which has been developed especially to simplify this operation. Lines and buses are coded by a starting x,y coordinate and length. Electrical symbols are added by x,y coordinate, length, format and data item

SCADA CONTROL CENTER FUNCTIONS

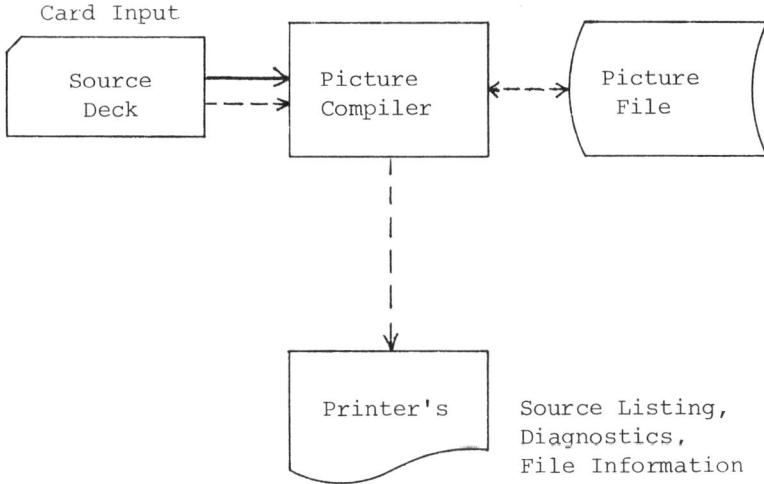

Fig. 10 Card Input Picture Compiler

    (data base) name. Electrical symbols, lines, buses, etc., whose color is dynamic, are coded to include data base names, for status information, similar to dynamic values.
c. Cards are punched from the coding forms produced from b.
d. The cards are processed by the Picture compiler which builds the Picture File.
e. Source listings, diagnostics, picture file, and compiler information are printed on the line printer.

    The Interactive Picture Compiler provides the facility for creating or modifying a CRT picture and the corresponding Picture Frame File. This is done from the consoles through the Data Entry facility as shown in Figure 11.

    The Interactive Picture Compiler (IPC) is activated by function button from the operators console. The IPC responds with a message to the operator telling him what to enter via his keyboard and how to proceed. (Note: the top 3 lines of each picture are reserved for messages to the operator and entry of information to the Control Center System by the operator.) After each entry by the operator, the IPC checks the entry for errors responds to the operator with a message asking for the next entry, and updates the picture file with the new information. The changes in the picture are displayed on the CRT as they are made. The operator proceeds in this manner interacting with the IPC until he has completed the addition/modifications to the picture. For dynamic values and

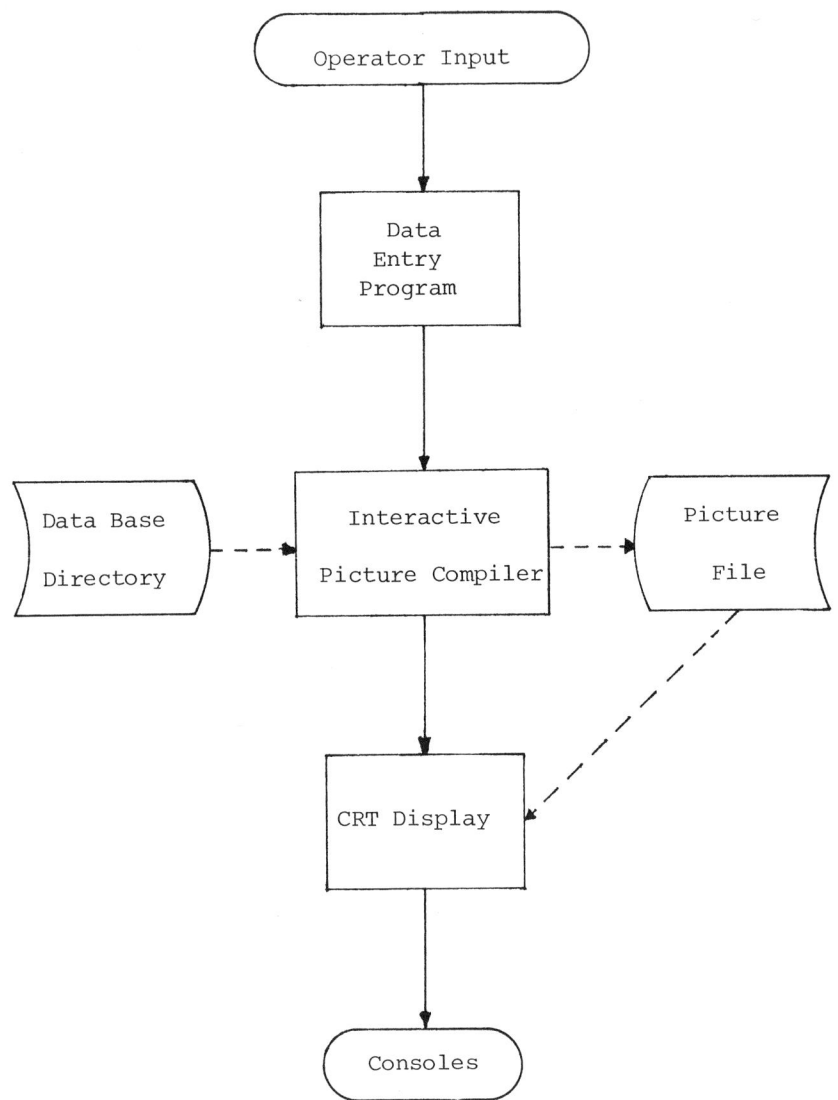

Fig. 11 Interactive Picture Compiler

status, line 2 is used to enter Data Base name and format information to establish the linkage with the Data Base through its Directory for dynamic update. The IPC is easy for the operator to use, provides instant feedback on the entries he has made, and the response-messages serve as a check to insure that all the necessary information has been entered for the creation/modification of a picture frame file.

The main disadvantage is that a source listing is not available for pictures created by the IPC.

The Picture Compiler provides the capability of implementing a CRT Picture as a main, future, or backup picture. Hence, new pictures can be implemented as a "future" frame and can be checked out before implementing them in the "on-line" system. Older versions of pictures can be saved as backup in case of software problems.

V. OPERATING SYSTEM

A. Requirements

The Operating System is required to provide a real-time priority oriented processing environment for the computer complex shown in Figure 2. It is also called the Executive.

The following principal capabilities and facilities are required of the Operating System:

    a. Multicomputing and multiprogramming environment for all program execution. A multiprocessor is highly desirable from the availability point of view.
    b. Real-time, on-line, demand, and batch scheduling of programs.
    c. Timer facility
    d. Interrupt handlers*
    e. Input/Output handlers*
    f. Facilities for error detection, error handling, failover, recovery, and automatic restart (described in Section VI)
    g. On-line file management
    h. Core management - dynamic core allocation
    i. Language processors
    j. Utility routines
    k. Library routines

B. Implementation

For state-of-the-art SCADA Systems and Control Centers, the Operating System (OS) is normally implemented by revisions

---

*For standard peripherals and non-standard devices which include the Man-Machine subsystem and the Master Station subsystem (Data Acquisition).

and add-ons to the latest version of OS for the computer subsystem (CS) selected. The primary reason for this approach is the prohibitive software cost of developing and maintaining a new OS optimized for the SCADA system.

The Operating System is designed to be an integrated software system comprised of the following functional modules:

  a. Program/Task Scheduler - builds the schedule queue for programs (jobs) according to job priority.
  b. Facilities Allocator - assigns processing and I/O facilities. For example, user core storage is assigned to a job based on space available. Real-time activities are assigned core residence until execution terminates. Demand tasks are assigned core storage on a lower priority basis and may be swapped out to mass storage when higher priority tasks enter the system. When core space becomes available, the lower priority task is rolled in (and relocated) and execution continues.
  c. CPU Time Allocator - assigns CPU time to the various activities within tasks, based on their priorities.

1. Program Scheduling

Programs, subprograms and routines are formed into executable tasks and scheduled for processing according to their priorities. At least the following priorty levels are provided (in descending order):

  a. Executive and interrupt processing
  b. Real-Time - these activities require fast response; e.g. Data Acquisition processing, alarming, and supervisory control.
  c. On-line/Demand - these tasks are executed on demand (operator console requests) or scheduled via a timer. They are time dependent but not as time-critical as real-time; e.g. CRT Display requests and data entries.
  d. Batch - provision is made for normal batch processing and for execution completion in a specified deadline time (CPU or elapsed time).

On-line scheduling and routing control facilities provide an effective mechanism for one program to call another into execution and pass it data. The scheduling packet is built and queued for execution. Interfacing facilities with the data base and on-line files are provided.

SCADA CONTROL CENTER FUNCTIONS

2. File Management

In addition to all the standard file management facilities for batch programs, file management facilities are provided for on-line programs to minimize CPU processing overhead. The following facilities are generally provided:

   a. Files are created, assigned, and modified in batch priority level. Address resolution is provided for all data transfers to mass storage and for file editing.
   b. Provision shall be made for the editing of files from an Operators Console.
   c. A logical addressing scheme is provided to facilitate word addressing of mass storage. File management maintains the relationship between logical and physical addressing.
   d. Dual storage of files is provided for high availability.
   e. File Management provides for and supports at least the following types of files:
      1) Permanent files with both fixed and variable length records.
      2) Temporary files which are maintained until released by the user.
      3) Scratch files which must be released by the user prior to termination.
   f. File Management provides Logical File control elements with capabilities for implementing:
      1) indexed files
      2) sequential retrieval of records
      3) partial key searches
      4) multiple keys for the same record
      5) multiple indexed files
      6) master index for indexed files

Logical file control is designed to minimize the number of mass storage accesses and the access time required to retrieve records.

3. Interrupt and Input-Output Handling

Provision shall be made for servicing all interrupts at Executive priority levels. These include:

   a. External interrupts generated by standard and non-standard hardware.
      1) Operator console requests through the Display

Generators.
2) Data Acquisition interrupts to include provision for Intelligent RTU's.
3) Completion responses, error responses, device time-outs, and non-response time-outs.
4) Foreign computer requests and interrupts.
b. Internal interrupts generated by both hardware and software.
1) Power failure
2) Parity (memory and I/O channel)
3) Protect violations
4) Internal rejects.

The <u>Data Base Handler</u>, which is part of the Data Base Manager, provides facilities similar to File Control for the main storage. It provides security against Data Base access by unauthorized programs, and a lock mechanism to prevent programs from reading data while another program is writing into the data base. Provision has been made for recovering the data base following a failure and restoring it prior to the initiation of application programs.

The <u>Timer Complex</u> provides a means for scheduling on-line programs or activating real-time programs by timer entry calls (primitives). Programs can be scheduled on a cyclic basis, or an absolute time of day or relative time can be specified for starting a cyclic timer (limited to 24 hours). Timer entries can be created, modified, deleted, enabled and disabled via Data Entry Form from the operator consoles. Programs can also be scheduled cyclically a finite number of times and with a given start time, and the scheduling can be repeated on a 24 hour basis.

Following a system failure, provision is made for recovering Timer entries and automatically adjusting timer counts to insure that scheduling "catch-up" is performed. Scheduling accuracy is maintained within 200 microseconds for real-time programs and 1 millisecond for on-line programs.

The <u>Master Station Handler</u> (<u>MSH</u>) provides a hardware and software interface with the Master Station subsystem. Similarly, the <u>Man-Machine Handler</u> (<u>MMH</u>) provides a hardware and software interface with the Man-Machine subsystem. These handlers provide features which simplify the user's communication with those systems. Primitives are provided within the OS for the convenience of the user calling these handler facilities.

VI. PERFORMANCE MONITORING

The Performance Monitoring function is required to enhance the operational availability of the SCADA or Control Center System. A basic requirement (and a design goal) is that the availability of the SCADA System and its communication subsystem and RTUs must be greater than the corresponding availability of the power system network. That is, to be of value, the SCADA system must remain in operation during power system outages, faults, and failures caused by electrical and mechanical malfunctions or hostile environmental conditions.

The basic philosophy of Performance Monitoring is that any single fault within the SCADA system shall neither "stop" the system from operating nor disturb the power system. Facilities are provided within the CS to failover to backup (alternate) hardware subsystems and devices, and/or to abort and bypass faulty programs such that recovery to operational status is ensured. The tools and facilities discussed in this section provide many paths for recovery. Hence, operation in a degraded mode is a procedural problem, not a technical problem, as the facilities for recovery have been provided.

Hardware-wise, availability of the system is increased by equipment redundancy; i.e., dual paths via communication loops, redundant computers, 2-port memories, dual mass-storage subsystems, redundant consoles and CRTs, etc. as shown in Figure 1.

Software-wise, availability is increased by the Performance Monitoring function, which is defined to include error detection, error handling, failover, recovery and automatic restart. Performance Monitoring is included in all of the SCADA and control center functions and covers both the software operations and control of the hardware.

From an implementation viewpoint, the Performance Monitoring function is distributed throughout the software. <u>Error detection</u> and portions of error handling are implemented as an integral part of each software facility. Provision is made for detecting errors at all levels of SCADA system operation. Calls to the Operating System (OS), Data Acquisition, Data Base, and Man-Machine software facilities are implemented with an error status returned to the calling program, and linkages to facilities which will perform corrective action to an error condition are provided. For most system operations using the OS, the system software facilities and the external hardware subsystems, a positive completion response is returned and verified for correctness on all commands and data transfers.

Automatic and Manual System Exercises are provided to periodically (and manually) activate all devices and paths to devices to determine their state of operability. Error checking is provided on all data transfers both internal and external to the CS.

For <u>error handling</u>, special software is developed which attempts to recover from detected error conditions before resorting to a failover to the backup system or unit. Error flags are set and alarm messages describing the error are generated for the computer operator and logged. For each detectable error, provision is made for handling the error by taking the appropriate corrective action which may be:

    1. Retransmission of command and/or data (3 times) upon detection of a data transmission error or the "re-try" of a function, program, or routine.
    2. Initiation of failover to backup device, program or data set upon failure to achieve a successful retransmission or retry of function.
    3. Aborting the function that is the source of error and releasing the facilities without interrupting the operation if backup device, program or data set is not available.

<u>Failover</u> is provided upon detection of a device failure or unrecoverable error. Provision is made for switching over automatically to a backup device, program or data set, for all single point (first contingency) failures associated with the critical function. Failover capability must be sufficient to enable normal operation to be resumed following the recovery and restart operations.

<u>Recovery and Restart</u> are provided following failover of the primary system or a primary device to a backup device. Provision is made for automatically reloading all required programs and data files, initializing the system, the restarting the processing for normal operation. Time corrections are made, queues rebuilt and facilities shall be reallocated as required. The extent of reloading and initializing to be done is dependent upon the failure. The Availability software is an integral part of the Exec. software package to satisfy these requirements and it must contain sufficient detection, decision, and control logic to effect failover and recovery for all single point (first contingency) failures.

VII. REFERENCES

(1). DyLiacco, T.E., "System Control Center Design" presented at the Systems Engineering for Power: Status and Prospects Conference, Henniker, New Hampshire, August 17-22, 1975.

(2). DyLiacco, T.E., "An Overview of Power System Control Centers" IEEE Tutorial Course on Energy Control Center Design, IEEE #77TU0010-9-PWR, July, 1977.

(3). Anderson, M.D. and R. A. Smith, "Approaches to Intelligent RTUs - New Benefits for Data Acquisition", IEEE PES Winter Meeting, New York, January 30, 1977, Paper #A77 087-0.

(4). Schlaepfer, F., D. Stewart, and J.A. Jordan, "Data Bases in Power Systems Operations and Planning". Presented at the Systems Engineering for Power: Status and Prospects Conference, Henniker, New Hampshire, August 17-22, 1975.

(5). Anderson, M.D. and R. A. Smith "Data Bases for Energy Management Systems - A Real-Time Operational Approach" 1977 Power Industry Computer Applications Conference Proceedings, May 25, 1977, p. 74-82.

(6). Hennington, W.B., T.W. Landers, and S.H. Bouchey, "An Interactive Energy System Data Base Compiler", C74365-3, IEEE Summer Power Meeting, Anaheim, Calif., July 1974.

Power Control System and Protection

LOAD MANAGEMENT - A SYSTEMS CONCEPT

Julius Orban

Sangamo Weston, Inc.

I. INTRODUCTION

In recent years a great deal has been written and said about load management. In the long term it is viewed by many as one of the key factors in reducing the nation's dependence on imported oil and conserving scarce natural gas that is currently used to generate peaking power.

This paper will provide a perspective on load management by discussing past experiences, current status, and future potential. System configuration and operating characteristics of the system components of an audio-frequency power line carrier load management system will also be discussed.

II. WHAT IS LOAD MANAGEMENT?

It is a systems concept of modifying the pattern of electric energy use in order to improve the efficiency of the electric energy system. Typically, load management is viewed by many in terms of its ability to shave system peaks for improved load factor. Load factors, both daily and annual, are cited as a measure of the effectiveness of load management and the reason for or against the implementation of such a program. But load management is much more than that. While it is true that load factors are convenient indicators, since they are usually known or easily determinable, it should not be used as the sole criterion for determining the need for load management.

Load management may be used to:

- Increase or decrease electric energy consumption at specified times.

- Reduce or balance loads on marginal substations and circuits, thus extending their life.

- Replace or augment reserve capacity, thus saving scarce fuel and capital.

- Provide extra margin in the event of forced outages.

A. Forms of Load Management

Load management falls into two major categories:

. Utility (supply) load management.
. Consumer (end use) load management.

1. Supply Management

The electric utility industry in the United States has traditionally practiced supply management. This form of load management includes:

- The design of the generation mix and capacity to meet the consumer demand in an optimal fashion.

- Interconnections with other electric utilities to exploit differences in load patterns.

- Storage of electric energy at the utility level (pump storage).

The economics of scale in generation, and reasonable fuel cost allowed the electric utilities to provide reliable service at low cost until the late 1960's. Since then, the concept of supply management has run into some major road blocks. Inflation, cost of capital, and increased fuel costs forced the per unit cost upward (as shown in Figure 1).

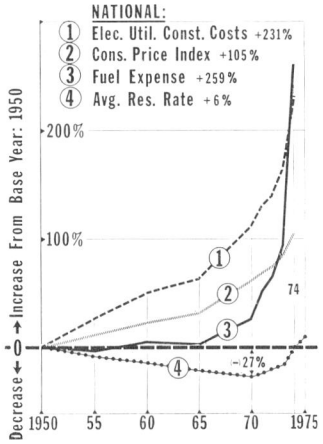

Fig. 1.

Environmental restrictions, siting and licensing delays forced the postponement, indeed the cancellation, of many new generating plants. The uncertain availability of the fuels typically used for peak generation (gas, oil) has further complicated the utility system planner's problem.

It is becoming more and more apparent that the traditional forms of utility load management (supply management) will have to be augmented with increased consumer load management.

2. Consumer Load Management

Consumer load management can be classified into two major categories:

- Indirect Load Management - Where the control of the end use loads is left to the customer.
- Direct Load Management - Where the utility controls the end use loads.

The goal of both of these techniques is to modify the consumption patterns of the consumer for more efficient operation of the electric utility system. The various consumer load management techniques include:

- Rate incentives to the consumer
- Energy storage at the consumer's level
- Promotion of off-peak loads
- Control of deferrable loads

The application of these techniques can further be classified as:

- Demand Management - Where the purpose is the reduction of peak demands, routinely or in an emergency, thus reducing peak generation requirements.
- Energy Management - Where the purpose is the modification of consumption patterns either to conserve energy, or to promote off-peak energy use for improved load factor.

III. LOAD MANAGEMENT - HISTORICAL PERSPECTIVE

Load management has been practiced for many years. In addition to Europe, Australia, New Zealand, to a limited degree even in the United States. In the U.S., some utilities had interruptable rates and time switch controlled water heaters for several decades. However, the control of these water heaters and the accompanying dual-rate structures (typically on and off-peak KWH) have been gradually phased out. The reasons for this were multi-fold. The increasing maintenance costs, reliability of the switching, the inflexibility of the control device coupled with the declining marginal operating and capacity cost, were all major contributing factors to the steady decline of load management. Further, regulatory pressures to reduce the rate differential between the on-peak and off-peak periods, reduced the economic incentive for the consumer and fewer and fewer opted for control. (Apparently many regulatory agencies viewed this form of rate structure as promotional.) As an end result, by the late 1960's and early 1970's, the controlled water heaters were virtually eliminated. (One notable exception was the Detroit Edison Company.) The interruptable rates were seldom, if ever, used since the supply management techniques by the utilities provided adequate supply to meet the demand.

Over this same time period the utilities in Europe, New Zealand, and Australia have increased their load management activities. As a result of the increased activity level, new technologies were developed. Load management has evolved from time switches to centralized load management systems. These systems provided greater flexibility, reduced maintenance cost and they have, with few exceptions, replaced time switches. The most prevalent system, used in Europe, is the ripple system. (This system will be described later in this paper.) Figure 2 shows the growth rate of the load management (ripple) systems by Schlumberger, the parent company of Sangamo Weston. Today there are some 915 systems operating over 3,000,000 receivers by Compteurs Schlumberger alone (see Figure 2). The total number of worldwide ripple installations are in excess of 1500, with some 6,000,000 receivers. It is interesting to note that in this time period the economic conditions were the same for these utilities as for their counterparts in the U.S. That is, marginal operating and capacity costs were declining, and siting and licensing delays were non-existent. Why did the European utilities opt for load management? This question is a complex one, based on policy decisions and beyond the scope of this paper. The most often cited answer to this question was that load management increased energy sales without the need for expanding the generating and distribution facilities.

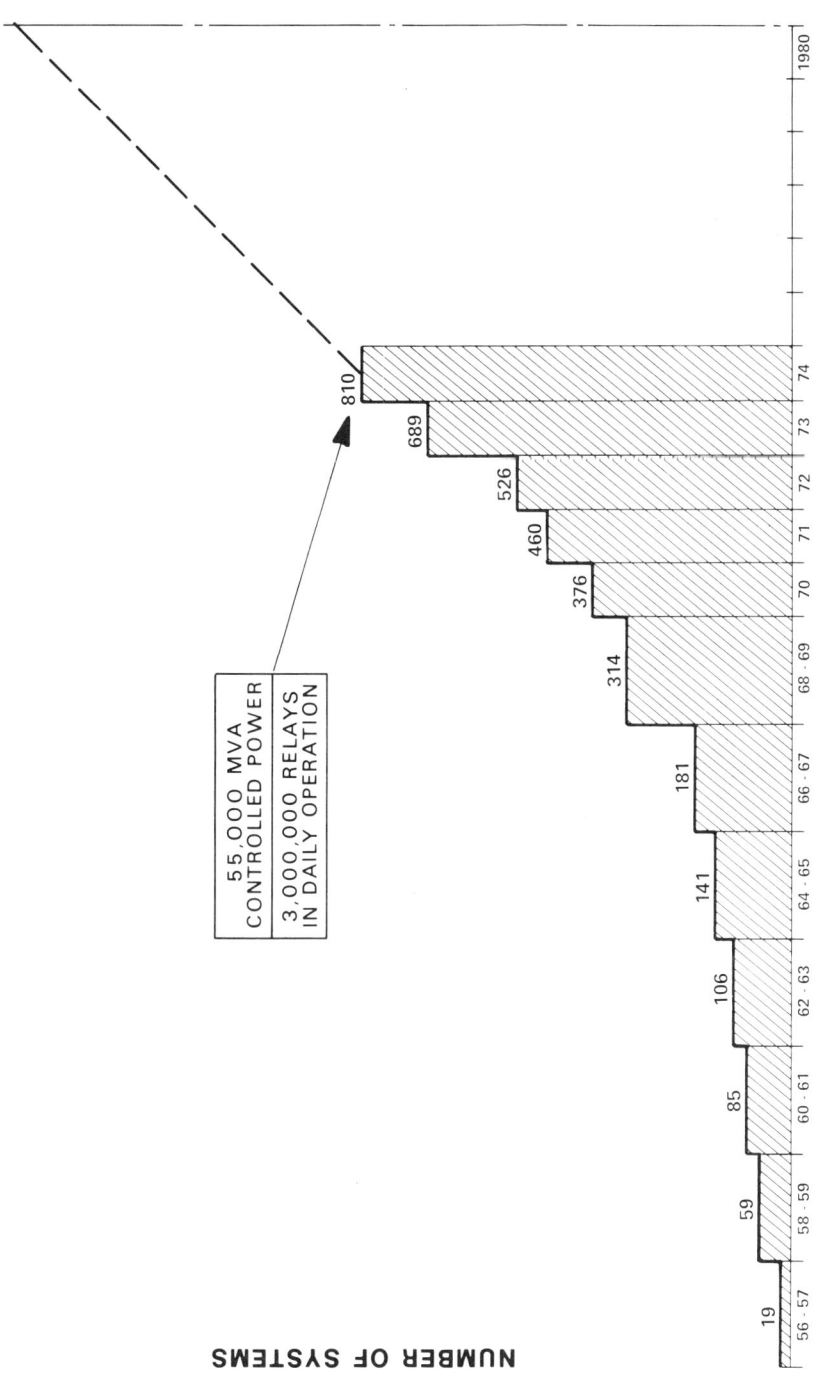

Fig. 2. GROWTH OF LOAD MANAGEMENT

A. Objectives of the European Utilities

While the economic climate is different today, both in the U.S. and Europe, than it was in the 1960's and the reasons for load management differ somewhat, the objectives of load management remain the same, that is:

- Demand Management
- Energy Management

In Europe, just as in the United States, different utilities were faced with different problems and the goals of their load management program varied. In areas where the problems were both capacity and energy related, such as the countries with hydro generation, the primary goal was demand management, the peak shaving of existing loads. In other areas, where the problem was capacity, the primary goal was valley filling with the introduction of new off-peak loads through storage devices.

The new off-peak loads enabled the utilities to increase total energy sales without adding additional capacity and it kept the base and intermediate generators on line even during the off-peak periods. This, in addition to reducing the start-up and shut-down costs, reduced the maintenance requirements by reducing the thermal stresses.

Peak shaving, through the direct control of deferrable loads (primarily water heaters), reduced the system peaks and enabled the connection of new loads which, in turn, maintained the revenues.

IV. CURRENT STATUS IN THE U.S.

A. Impact of Consumer Load Management

One of the main concerns of the U.S. utilities regarding consumer load management is the unavailability of data relating to the long term effect on operating cost and reliability of the system. The European utilities limited their analysis to the short term costs and benefits. As a result, the marginal production costs, caused by the change in load patterns and the subsequent changes in generation mix, were not included.

Studies such as the ones by the Edison Electric Institute's Planning Committee (1), and the Systems Control Inc. (2), have further clouded the issue.

1. The EEI Study - Phase 1

This study (1), in its conclusions, warns that: The benefits from higher annual load factors might be less than the cost of implementing the load management program for improving load factors. On the question of reliability, this same report states that improved annual load factors would require the increase in generation reserve for the same degree of reliability. The study was based on a large "predominately oil-fueled utility" and it investigated six different load patterns, using a fixed set of generation resources.

2. The Systems Control Inc. Study

This study (2), simulated the effects of load management on two U.S. utilities. One, a primary coal-fired, and two, a primary oil-fired system. Both systems also had nuclear plants and burned natural gas.

This study concludes that energy management resulted in large savings for the oil-fired utility and demand management reduced the operating costs for the coal-fired utility. Figure 3 shows the average fuel cost versus annual load factor. The solid lines represent constant peak demand as the energy supply is changed and the dashed lines represent constant energy sales as the peak demand changes.

It can be seen that for an oil generating utility, the slope of the changing energy and constant peak (solid lines), is greater than for a coal generating utility. This implies that the fuel costs are affected more by energy management. At the same time, the slope of the changing peak and constant energy (dashed lines), is greater for a coal generating utility, implying that this utility will gain greater benefits from demand management. Figure 3 also shows the limit to the operating benefits gained from load factor improvement, since the average fuel costs tend to level off at load factors over 65%. The dotted lines represent the constant net operating revenue.

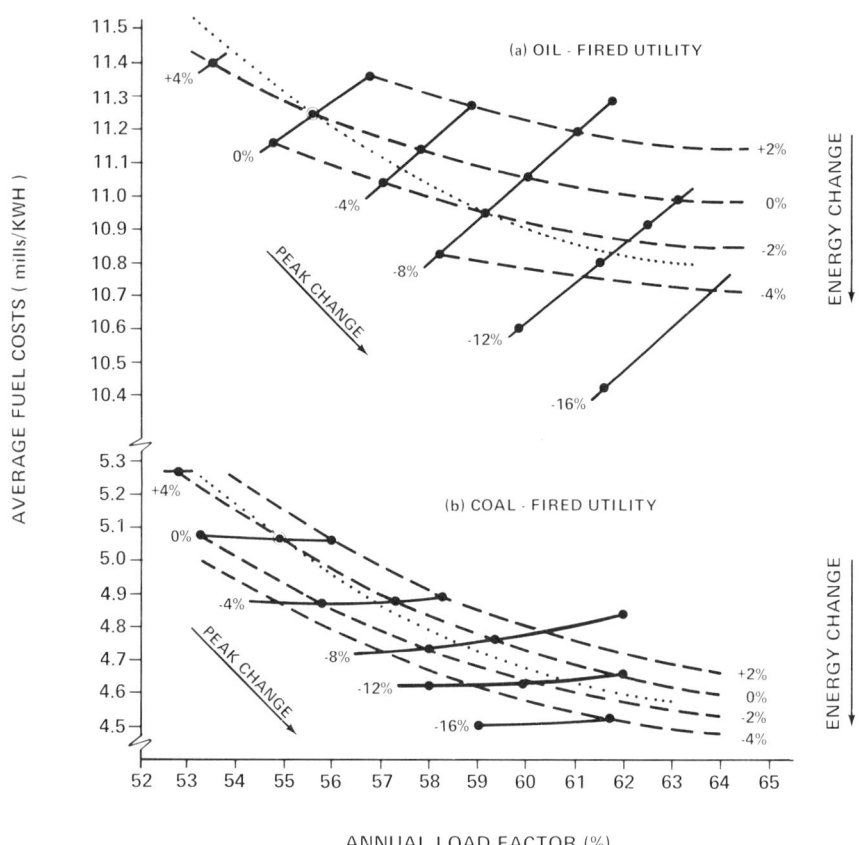

Fig. 3. AVERAGE FUEL COST vs ANNUAL LOAD FACTOR

LOAD MANAGEMENT—A SYSTEMS CONCEPT 71

3. Reliability and Maintenance

On the question of reliability, the report finds that load management could improve the system reliability. Figure 4 shows the loss-of-load probability versus the peak demand. It can be seen that a 1% decrease in peak demand decreased the loss-of-load probability approximately 20%.

The Systems Control Inc. study found that for both utilities, the generation was sufficient to perform maintenance even if daily load factors were 99%.

4. Why the Contradiction?

The results of the two reports, by well respected organizations, appear to be contradictory. One report states that load management improves reliability - the other states just the opposite. One could argue that EEI omitted a very important factor from their reliability consideration when they failed to include load management as part of the overall reserve capacity. Certainly, controllable loads can be treated as part of the spinning reserves since these loads can be shed in the event of an emergency. On the other hand, one could argue that if load management is an integral part of the system, calculations of reliability must allow for possible failures in the control equipment.

Of course, the reason for the different conclusions can be attributed to the profiles of the utilities. Different generation mixes and load patterns will produce different cost/benefit results and, as such, load management should not be undertaken without a thorough study on the individual system.

V. Experiences In The U.S.

While each utility must perform its own analysis to determine the effects of load management on its system, the experiences of some of the utilities are worth reviewing.

A. Indirect Load Management

1. Connecticut Peak Load Pricing Test [3]

This study is the first of several FEA funded projects in the area of indirect load management. It was conducted in the Connecticut Light & Power Company service area from October 1975 to October 1976, on a stratified sample of residential customers (184 customers completed the test).

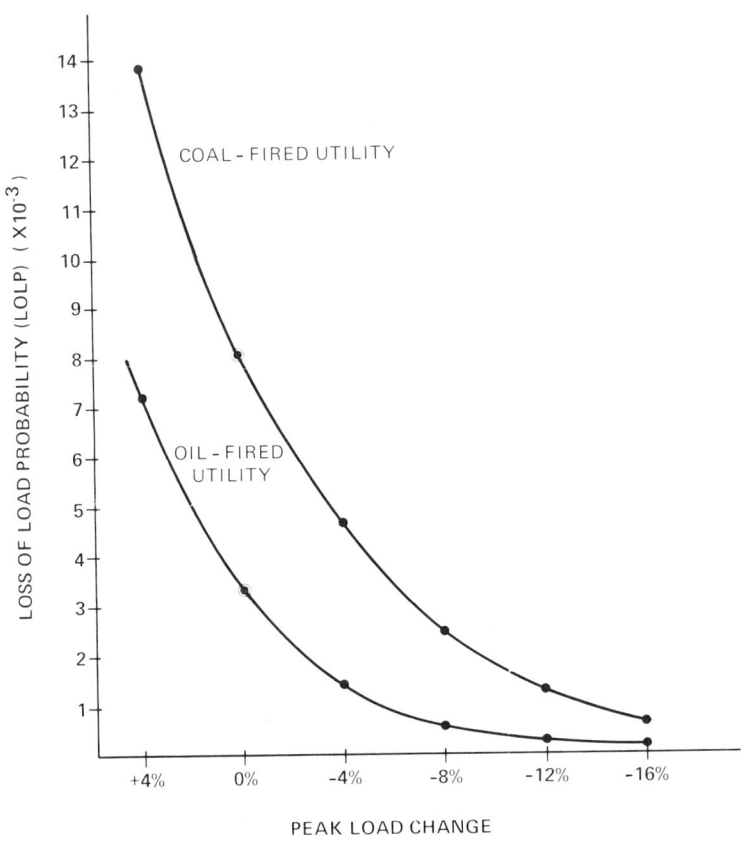

Fig. 4. LOSS OF LOAD Probability vs PEAK CHANGE (at constant energy)

Rate Structure:

    16¢ per kWh - On-Peak Period
    3¢ per kWh - Intermediate Period
    1¢ per kWh - Off-Peak Period
    Plus - $2 Monthly Customer Charge

The prices were derived from the cost studies performed by the utility.

Time intervals:

    Peak Hours: 9:00 a.m. to 11:00 a.m.) Winter
                5:00 p.m. to 7:00 p.m.)

                10:00 a.m. to 12:00 Noon) Summer
                1:00 p.m. to 3:00 p.m.)

    Off-Peak Hours: 9:00 p.m. to 8:00 a.m.) Both Seasons

    Intermediate Hours: All remaining.

2. Conclusions

The consumers in the test group had:

    a. Shifted substantial amounts of consumption out of the peak and intermediate periods into the off-peak hours. This was true for both small and large users of electricity. The difference in consumption used on-peak was 23.1%.

    b. Higher load factors. Based on coincident demand with system peak, the improvement in load factor was 14%.

    c. Altered schedules rather than use. This was true even on days of extreme temperature.

    d. Approved the innovative rate structure and a large majority were willing to continue in the test.

B. Direct Load Management

1. Water Heater Control

    a. Detroit Edison - Has about 200,000 controlled water heaters on the system and is capable of shaving an estimated 180 MW or about 0.9 kW per water heater.

b. Buckeye Power - Will have 40,000 controlled water heaters. Tests run during the 1974-1975 winter show an average 1.4 kW of diversified demand per water heater can be shed.

2. Air Conditioner Control

   a. Georgia Power - Of the 300 residential customers tested, the results show a 1.4 kW reduction per customer. This means a 25% reduction in the demand with "no change in kWh" consumption, or a 33% improvement in load factor.

   b. Arkansas Power & Light - The peak coincident demand dropped per air conditioner was from 4.08 kW to 4.17 kW with "little change in kWh use."

   c. Detroit Edison - A 25% reduction was achieved with "little change in kWH." At 95°F temperature, the estimated emergency load shedding potential is 800 MW.

3. Electric Space Heating Control

   a. Potomac Edison - The control of space heating, water heaters and clothes dryers defer a minimum of 8 kW per customer from the morning and evening peak. Less than 4 kWH per customer was lost. Total potential: 210 MW.

4. Irrigation Control

   a. Southwest Public Power District - Estimates the saving of $500-$1000 for each 100 H.P. motor in operation for 1000 hours by "scheduling" irrigation pumps.

5. Storage

   In addition to load control, the obvious method of load deferral is through the use of storage devices - both for space heating and cooling. A study by Dairyland Power Cooperative shows the following:

   The diversified demand for the average home at the time of system peak is 6.3 kW for baseboard and ceiling cable and 3.5 kW with stored heat by off-peak floor cable. It is 15.9 kW for electric furnace and 9.9 kW for heat pump.

   Further, a cost of service comparison shows a 33% saving with storage heat.

   All of the utilities report good customer acceptance.

While the evaluations of load management are complex, the preliminary results from these studies indicate significant, short-term benefits. Of course, comprehensive economic analyses are necessary to determine the long-term effects of load management.

VI. COMMUNICATION TECHNIQUES FOR LOAD MANAGEMENT

There are four main techniques for load management and/or distribution automation:

- Telephone
- Dedicated Lines
- Radio
- Power Line Carrier

For the purposes of this paper, the Sangamo Weston Audio Frequency Power Line Carrier (Ripple) System will be described.

VII. THE OPUS 6000 LOAD MANAGEMENT SYSTEM

The Sangamo Weston OPUS 6000 is a real time centralized control system that uses the 60 Hz power distribution network as the propagation medium for control signals. Centralized real time control is achieved by transmitting audio frequency (175 Hz to 390 Hz) pulse codes from one or more injection points to receivers located throughout the network. The two basic modes of control are shown in Figures 5 and 6.

The coupling circuit components (Figure 6) include:

a. <u>A high Q series tuned element</u> - which limits the 60 Hz power absorbed by the coupling circuit and enables the coupling of the control message to the H.V. bus of the 60 Hz network.

b. <u>An injection transformer</u> - provides the designed injection voltage level (1 to 2%) by the appropriate selection of the turns ratio.

c. <u>An injection switch</u> - short circuits the primary windings of the transformer outside periods of message transmission and connects the output of the pulse code generator to the primary of the injection transformer, under control of the control console, during message transmission.

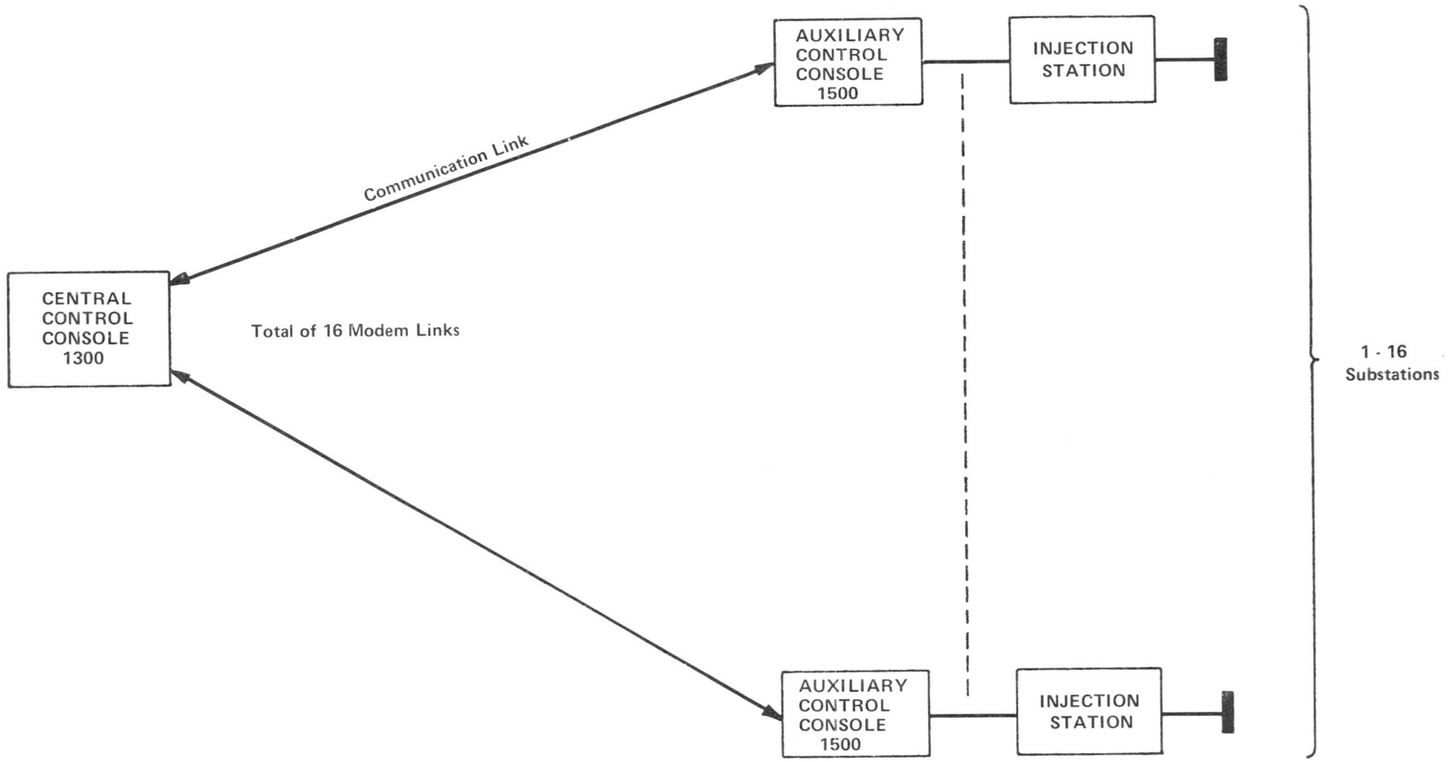

Fig. 5. Typical control of injection stations from one central point.

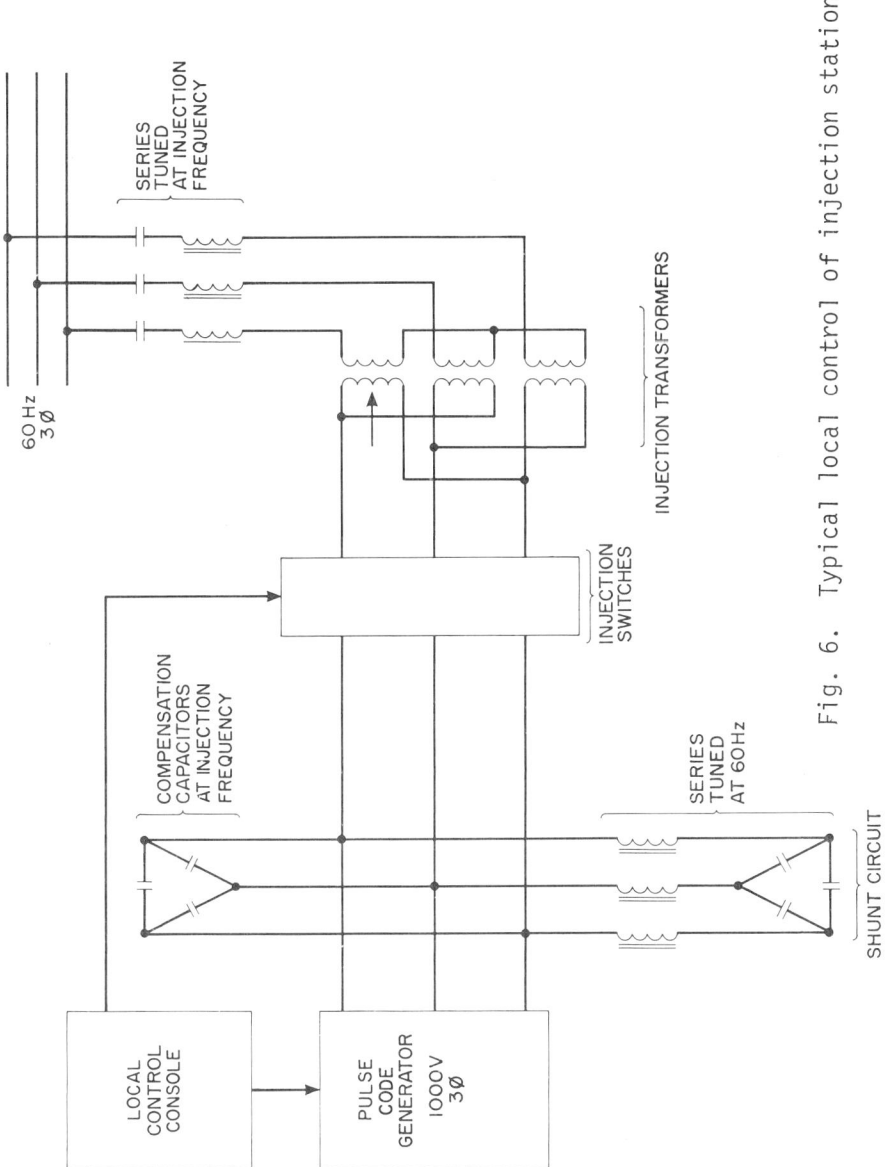

Fig. 6. Typical local control of injection station

d. A shunt circuit - consisting of a series tuned circuit at 60 Hz, which keeps the primary of the injection transformer shorted (for 60 Hz) during transmission periods. This portion of the shunt circuit, together with the injection switch, short circuits the injection transformer primary at 60 Hz at all times. In effect then, the transformer acts as a current transformer at 60 Hz and as a power transformer at the audio frequency, thus providing negligible loading to the 60 Hz network.

e. Compensation capacitors - compensate for the magnetizing current absorbed by the tuned circuits and the injection transformer at the audio frequency used.

Since the injected signal levels are low compared to the 60 Hz network voltage (1 to 2%), the protective relaying scheme of the utility need not be changed.

A. Message Code

A control message for the OPUS 6000 consists of a start bit followed by one or more execuite bits. Figure 7 shows the two available codes for the OPUS 6000. The 40-bit code operates the electromechanical and solid state receivers and the 31-bit code operates the solid state receivers. The timing of the execute bit in relationship to the start bit is precisely controlled because the receiver response to the control message is based on the timing between the start bit and the assigned execute bit. Since it is required to address and control many different groups of customers, throughout the network, the message code must provide several hundred discrete and mutually dependent commands. The 40-bit code allows 1305 unique commands and the 31-bit code 14,168. The code structure used depends on the control requirements and the design constraints that must be satisfied to produce an economical, dependable system.

B. System Complement and Control Functions

A centralized load management system should be safe, fast-acting, easy to operate and install, flexible enough to permit power distribution system expansion and have sufficient capability to control large loads. The OPUS 6000 Load Management System can use either a predetermined control pattern and/or operate in any arbitrary as-needed mode. The system, together with other Sangamo Weston products, can provide the utility industry with both direct and indirect load management capabilities. Figure 8 shows a fully expanded OPUS 6000 Load Management System.

40 - BIT CODE

STARTING PULSE

EXECUTE PULSES

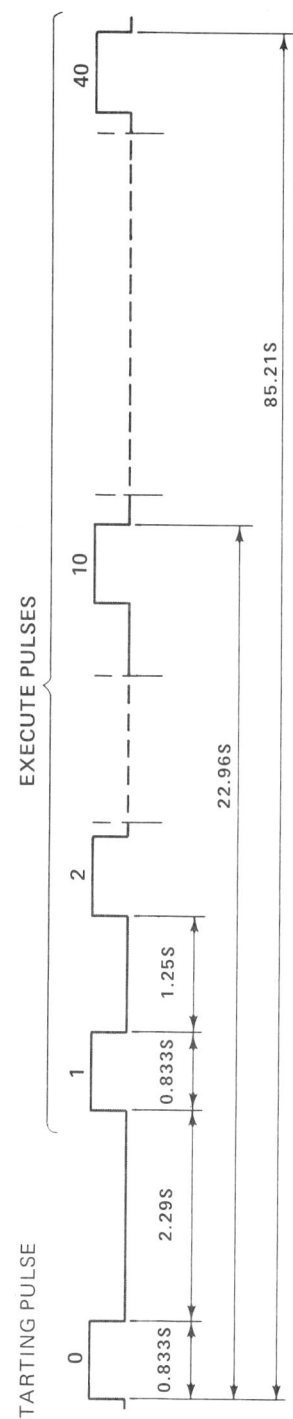

31 - BIT CODE

STARTING PULSE

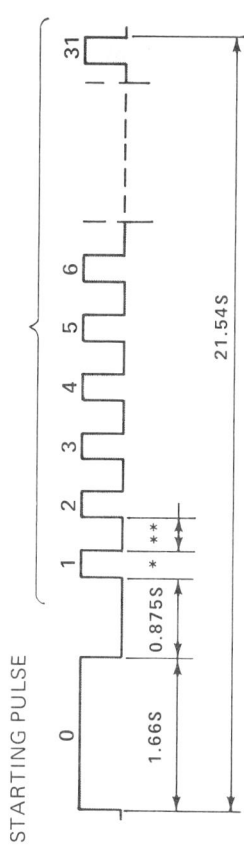

Fig. 7. OPUS 6000 CODE

S = SECONDS
* = 0.25S
** = 0.375S

The system can optimize operation of electric utilities by performing the following functions, either singly or in any combination, under automatic or manual control:

a. <u>Control of Time Dependent Load Patterns</u> - This approach to load management is based on fixed time daily programming to control the predictable daily, weekly, monthly load variations.

b. <u>Control of Capacity Dependent Load Patterns</u> - This refers to load management programs where certain loads may be interrupted for short periods to reduce peak loads or relieve loads on particular circuits.

c. <u>Control of Weather Dependent Load Patterns</u> - This approach can be used to control the charging periods and time of storage (heat/cooling) dependent on outside temperatures. Also, the control period or cycle time of air conditioning and/or space heating may be made time dependent.

d. <u>Emergency Load Shedding</u> - This may be used to disconnect large blocks of loads to ensure system stability in an unpredicted crisis situation, such as loss of generation.

The modular construction of the OPUS 6000 System lends itself to system expansion without cost penalty. Thus, the utility may purchase the minimum equipment complement and expand this complement as needed in the future.

VII. CONCLUSION

A. Future Potential for Load Management

Recent trends in the utility industry and certain policy decisions by the Federal Government indicate that load management will play an increasingly more important role in the normal operation of the industry. The trends include:

- Great increases in the price of fuels, resulting in greater energy costs to the utility and the consumer.

- The environmental concerns.

- The great increases in utility construction costs.

- The growing awareness by the public of the use of energy.

- The technological improvements in the area of load management.

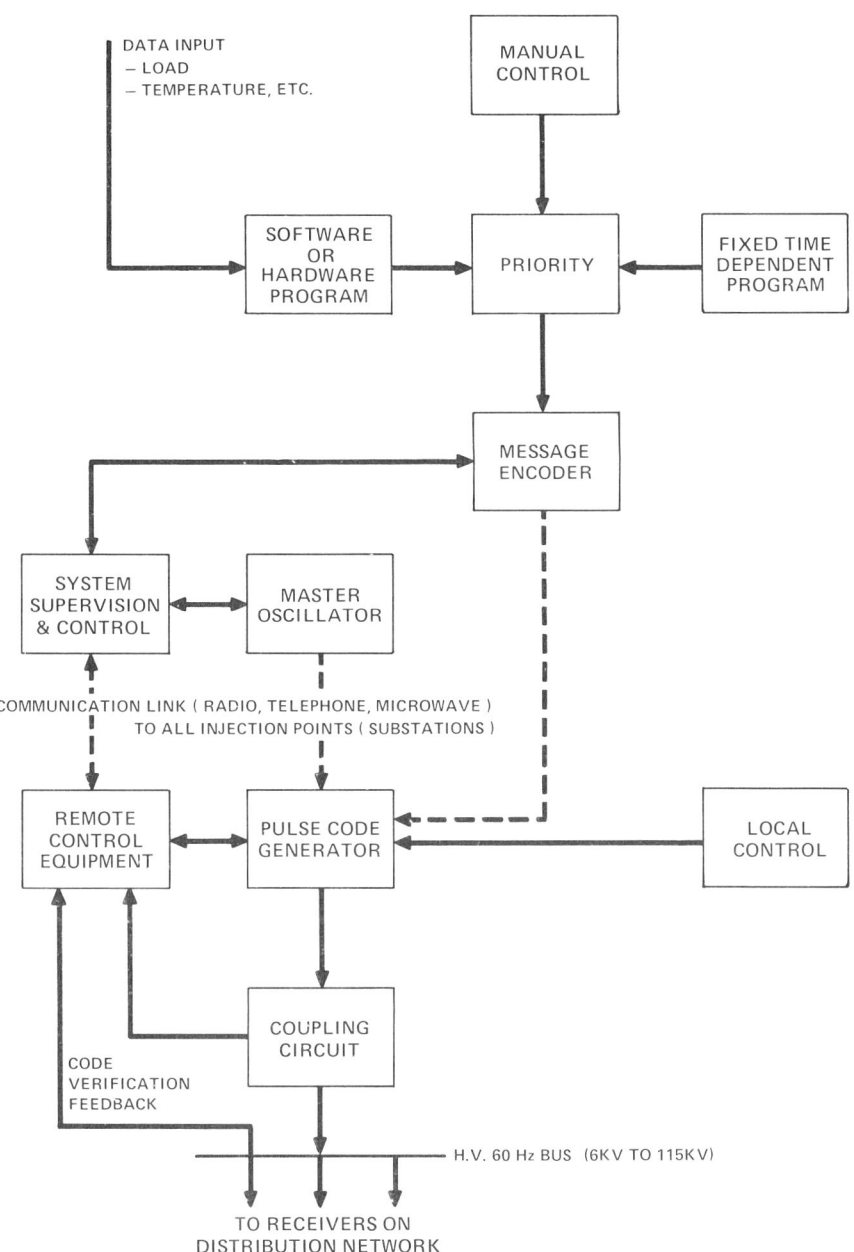

Fig. 8. OPUS 6000 BLOCK DIAGRAM

It is clear that load management expands the options available to the electric utilities. What is not clear is the rate at which load management will grow. Recent events, such as the U.S. Senate Energy Committee's action, which would force the U.S. utilities off natural gas; the President's Energy Plan to convert from oil to coal; the taxing of utility oil and gas burners; and mandatory time-of-day rates will all provide a greater impetus for load management. These events, coupled with the growing consumer awareness and their apparent acceptance of certain load management techniques indicate that in the not too distant future the concept of load management will become a routine part in the utility operation and system planning.

VIII. REFERENCES

(1). EEI System Planning Committee Report, "Load Management - Its Impact On System Planning And Operation, Phase 1". EEI Publication Number 76-28/April 76.
(2) Federal Energy Administration, Final Report to the Utility Program, "Analysis Of The Impact Of Energy Management Strategies Upon Electric Utility Costs And Fuel Consumption Patterns". Submitted by Systems Control, Inc., Palo Alto, California, September, 1976.
(3) Connecticut Public Utilities Control Authority
Connecticut Department of Planning and Energy Policy
Connecticut Office of Consumer Counsel
Northeast Utilities "Final Report Connecticut Peak Load Pricing Test" May 1977

Power Control System and Protection

THE DEVELOPMENT AND SELECTION OF ALGORITHMS FOR RELAYING OF TRANSMISSION LINES BY DIGITAL COMPUTER

J. G. Gilbert

Pennsylvania Power and Light Company

E. A. Udren

Westinghouse Electric Corporation

M. Sackin

Westinghouse Research Laboratories

The concept of transmission line relaying by digital computer has attracted much attention from protection engineers in recent years. Growing interest in the use of small processors for critical protection functions is evident from the number of technical papers describing study of and experimentation with such methods. While the relative economics for computer relying systems versus conventional solid state or electromechanical relay equivalents are not yet entirely clear and are changing there are definite technical and performance advantages which computers can bring to the relaying art.

Presented below is a brief summary of advantages for the digital approach; the historical background of methods for protection with computers and a more detailed description and analysis of the most useful specific methods recently tested by the authors. The testing procedure will be explained and the results reduced to a ranking of methods based on the authors' evaluation criteria.

I. ADVANTAGES OF COMPUTER RELAYING

Some benefits of computers for protection were immediately apparent when such applications were first seriously considered about ten years ago; others were discovered only after experimentation and experience. The key advantages are:
  1. Flexibility - a single, general purpose hardware base can be used to perform a variety of protection and control functions with change of the programming only. Several of these functions can reside in the

processor memory simultaneously and can be called upon as needed. Similarly, drastic alterations or additions to protection logics can be made in the field with little or no hardware replacement.

2. Adaptive capability - a processor can be programmed to automatically change its behavior depending upon external circumstances which change with time. The basis for the change can be either local information available directly to the processor, such as load flow in the protected apparatus; or the change can initiate from an external source of intelligence, such as a substation operator or a data link from a central system control computer. The change may only be in a specific setting, or a whole new protection routine can be selected when needed.

3. Data-interface access - General purpose digital computer systems can always be equipped with input/output ports through which data and control commands can be exchanged. For example, the sequence of software events which occurs in the processor in response to a fault can be stored during the fault and output to a data link afterwards. A control computer or operator is provided with detailed information on specifics of the fault and the resulting software action. In turn, data can be input to the processor memory to improve or modify program response as described in (2) above.

4. Self-checking ability - Conventional relays are idle for essentially their entire lives - faults are present for a total of, perhaps, only a few seconds out of 20 or 30 years of service. At other times, the ability of the equipment to do its job is in doubt and must be confirmed by periodic testing. A digital processor, on the other hand, is by nature a dynamic device, and most hardware failures are flagged as soon as they occur by a processor stop. Beyond this, specific programs can be executed during non-fault periods which test the processing hardware, the integrity of the program memory, the calibration of the analog to digital interface, and more. Even though a small digital processor may be no more reliable than a conventional relay system in the hardware-designer's sense, it may prove both more reliable and more secure in the relay engineer's sense because it can alert the user to a malfunction before a false trip or failure-to-trip occurs.

DEVELOPMENT AND SELECTION OF ALGORITHMS

5. Possible economic benefits - While the costs of conventional relays steadily increase, those of digital processors suitable for apparatus protection are plummeting. This trend is likely to continue; even now, the processor with all peripheral devices and packaging needed for line protection is priced in the same neighborhood as a compliment of conventional relays. The prospective user should remember, though, that software development costs are not declining and will dominate over hardware costs unless a large number of identical systems are built.

6. Detailed logical and mathematical capabilities - While the conventional relay designer is somewhat constrained by the characteristics and limitations of electromechanical elements or solid state sensing circuits, the relaying programmer is free to provide almost any function within the limits of his imagination or understanding. Specific protection problems can be broken down into fine details, and each handled separately. A simple example related to distance-type protection of transmission lines is the ease with which arbitrary operating-characteristic shapes in the R-X plane can be provided. In addition, measurement problems can be stated as mathematical equations and directly implemented.

Many investigators of computer relaying have recognized and proposed possible equations which can serve as bases for distance-type protection of transmission lines. The analysis and testing of these varied approaches is the principal subject of this article.

II. SUMMARY OF LINE PROTECTION BY COMPUTER

Although a broad spectrum of arrangements are possible for application of computers to relaying, the authors' interest centered around the simplified hardware configuration shown in Figure 1, as applied for distance protection of a single EHV transmission line. An analog input subsystem accepts three-phase ac quantities from power system transducers, such as conventional ct's and ccvt's. All of the quantities are sampled simultaneously at uniform time intervals, between four and 32 times per ac cycle, converted to digital form, and transferred to the digital processor. The processor stores, organizes, and makes decisions based on the

values of the samples. Its prime purpose is to output a
breaker-tripping command when a fault occurs on the protected
line. Other secondary control and data outputs may also be
provided. Stored in the processor memory is an elaborate
program which implements the line-protection sensing and
logic. At the core of the relaying program is an algorithm
or equation which operates on the incoming raw data samples
to provide meaningful indicators of fault presence or location.

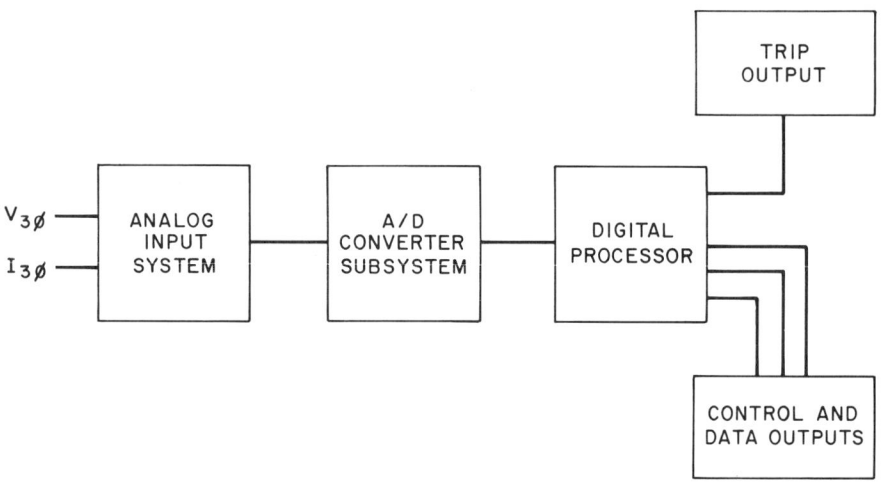

Fig. 1. Simplified Hardware Configuration

III. PROGRESS TO DATE

The first widely-recognized milestone in computer relaying is the 1968 paper by Rockefeller [1]. This work describes
in detail a hypothetical large-scale computer system which is
programmed to protect all of the apparatus in and radiating
from a major substation. Many specific programs are given
there. In 1970, Mann and Morrison described the derivation
and testing of programs for line protection [2], [3]. Their
work centers around an equation which yields the peak value
and phase position of a sinusoidal waveform from a group of
three arbitrary, consecutive, and uniformly-spaced samples
of the wave; the sampling need not be synchronized to the
power system frequency. Voltage and current peak values can
be divided to yield impedance magnitude from the relaying
point to the fault; the argument of the impedance is found by

simply subtracting V and I phase positions. The equation or
algorithm is derived by trigonometric manipulation of relations among values and derivatives of a pure sinusoid. It
can be viewed as a specific form of curve fit. This method
was a major breakthrough - the use of asynchronous sampling
removed the need for external synchronizing hardware and placed
virtually all of the relaying intelligence into software.
Around the same time, Westinghouse Electric Corp. and Pacific
Gas and Electric Co. (PG&E) developed an experimental minicomputer based system for non-pilot distance protection of a
single 230KV transmission line [4]. The experimental relaying processor, called Prodar 70, has been in service for six
years at PG&E's Tesla substation [5]. It provides three
zones of phase- and ground-distance protection, with high-set
overcurrent tripping, out-of-step blocking, reclosing, plus
a variety of logging functions. No false trips have occurred
in the field; the system has responded as expected to a large
number of external faults and disturbances, plus internal
faults, both natural and staged. The algorithm used in
Prodar 70 is similar to that proposed by Mann and Morrison,
with modifications to reduce errors from dc offsets.

The hardware used in Prodar 70 was about ten times as
expensive as the conventional static relays it replaced,
even if software development effort is disregarded. Because
of this cost disparity, only one additional experimental
system was reported during the next several years [6]. Many
other investigators, though, saw that the economic trend
favored the digital approach for the future, and proposed new
and ingenious algorithms and related processing techniques.
Ramamoorty [7], Phadke, et. al. [6], Carr and Jackson [8],
Laycock and McLaren [9], and others proposed various forms of
Fourier analysis, sometimes expressed as a waveform correlation. A key departure in these methods is that the fault
waveform is no longer assumed to be a sinusoid. Components
other than the 50 or 60 Hz fundamental are expected; the
correlation process acts like a notch filter. Horton [10],
Hope and Umahmaheswaran [11], and others described the notion
of correlation notch-filtering using reference waveforms
other than sinusoids - notably square waves.

Viewing the problem in yet a different way, McInnes and
Morrison [12] considered a simple R-L lumped-parameter model
of a transmission line. They developed, from the differential
equations describing that model, an algorithm which yields
R and L values from the relaying point to the fault directly.
Decaying dc offsets in the current are included in the solution and do not cause errors as in some of the other approaches described above. Ranjbar and Cory [13] suggested a

modification in which limits of integration are chosen to suppress specific undesired harmonics not accounted for in the lumped R-L model.

Finally, another group of papers proposed methods which can be loosely categorized as curve fits. Gilbert and Shovlin [14], and Makino and Miki [15], proposed fast-responding sinusoidal fits. Luckett, et. al. [16] briefly described a general least-squares fit.

The above listing of methods represents a rather complete sampling of techniques which have been proposed for post-fault steady-state impedance measurement used in transmission line protection. At the time of this writing, there is emerging interest in equations based on the traveling-wave behavior of lines; such methods are not analyzed or tested here.

IV. WESTINGHOUSE AND PP&L STUDY

Westinghouse Electric Corporation and Pennsylvania Power and Light Company (PP&L) have undertaken a joint study of computer relaying methods, with emphasis on techniques which appear useful as part of a commercially-feasible digital line protection system. This emphasis differs from that of the Prodar 70 project described above [4], which did not consider economics or other lesser practical aspects such as physical size or means of operator interface. Major components of the study are: investigations of line protection algorithms, evaluation of digital processors and peripheral devices appropriate to the relaying task, formulation of schemes for special hardware devices and interfaces, mapping of general relaying approaches and flowcharts, and comparison of alternative transducers of ac voltage and current data from the transmission line.

This article documents the results of a portion of the algorithm-testing phase of the study. The authors combed the literature in search of relaying algorithms of the types described above. A total of 68 papers which treated some aspect of protection or related control of power systems by computer were reviewed. Nearly half of these contained some explicit algorithm or method, but many were closely related or identical as indicated above. The investigators extracted 14 specific algorithms for subsequent exhaustive testing. The tests yielded a large bulk of results which were reduced with decision matrices described later.

V. PURPOSE OF ANALYSIS AND TESTING

The summary of the development of algorithms given above pointed out that each of them is based on specific assumptions of the nature of the ac waveforms being processed. One which is common to all the methods described here, as well as to virtually all conventional relays, is that buried in the transient-laden signals available immediately after fault inception are meaningful steady-state quantities, such as complex impedance, which are useful in locating faults. Beyond this, the methods diverge in their approaches. The authors have divided the 14 algorithms tested into four broad categories of Fourier and Walsh Analyses, Solutions of Differential Equations, Sample-and-Derivative Calculations, and General Curve Fits. Any of the methods is perfectly accurate when the assumptions from which it is generated are strictly observed by the ac data. However, since power system faults produce complex transient waveforms which cannot be perfectly characterized in advance, the results of algorithm processing will contain errors. Algorithms based on gross and simple assumptions, such as those presuming pure 50 or 60 Hz sinusoids respond quickly since they do not process enough data to establish what components are actually present. They generally have short data-windows, which means that only a few consecutive samples are needed to produce the output quantity. However, signals other than the presumed fundamental appear as noise and cause errors. Conversely, algorithms based on the presumption of complex waveforms generally must process a large number of data samples to extract the buried fundamental steady-state value. It takes longer to obtain results after fault inception with such a long data-window, but the filtering provided by the algorithm and/or auxiliary filtering routine gives greater accuracy for contaminated waveforms.

The object of testing is to subject each algorithm to simulated faults typifying service conditions anticipated by the authors for future computer relaying systems; and to find the algorithm which gives the best tradeoff of accuracy and speed for these faults.

VI. LISTING OF ALGORITHMS TESTED

The following gives the specific form of each algorithm tested, and briefly describes key characteristics. The reader should refer to the original source for greater detail. For each case a frequency-response magnitude plot is produced by performing a z-transform on the sampled-data expression of the algorithm [17], [18]. The plots in Figure 2 show the transfer functions of the algorithms, with frequency of the ac input waveform before sampling as the abscissas. The relationship between frequency and the complex variable z is such that the spectrum is folded about one-half the sampling frequency as predicted by the sampling theorem. In practical computer-relaying systems, the incoming analog data must be filtered to remove all components above one-half of the sampling frequency to prevent errors from the ambiguities of folding. Therefore, only the portions of the spectra below one-half the sampling frequency (also called the Nyquist limit) are shown. The reader should observe that those algorithms with long data-windows (i.e., those which process a long string of consecutive samples to yield an output) have frequency-response plots like those of sharp notch filters; such a filter is expected to respond slowly. Conversely, those with short windows act like broad filters whose output can change quickly but which may pass, or even emphasize, transient components above or below the power system frequency. Such components, in turn, produce transient errors in the output.

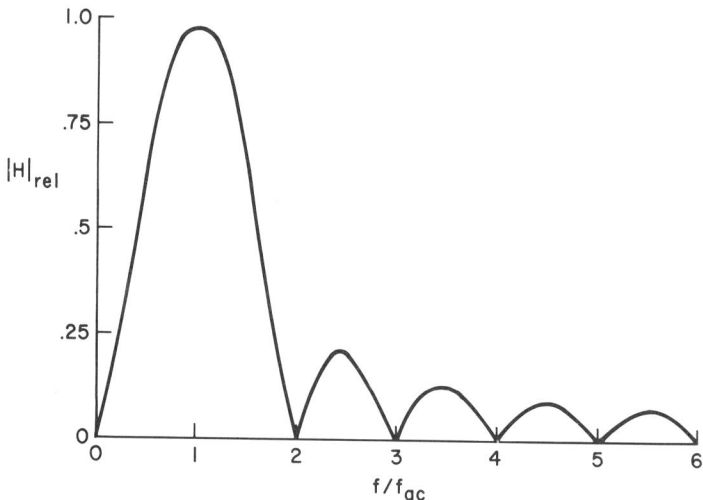

I. FOURIER ANALYSIS – ONE-CYCLE WINDOW

DEVELOPMENT AND SELECTION OF ALGORITHMS

2. FOURIER ANALYSIS – SHORT DATA WINDOW

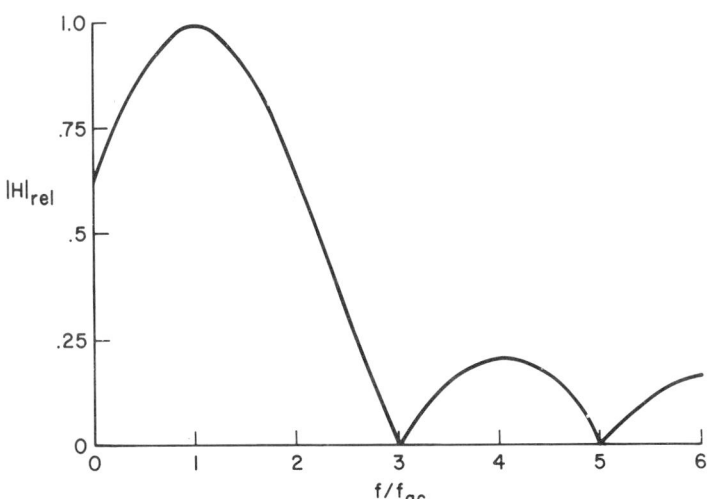

3. WALSH ANALYSIS

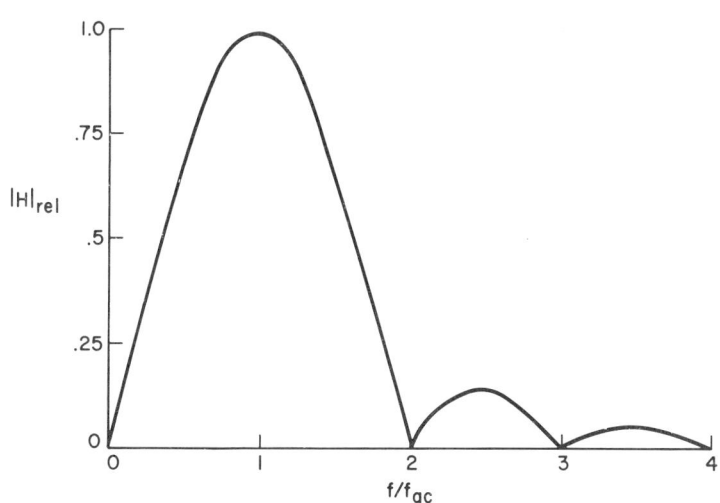

4. SAMPLE-AND-DERIVATIVE ALGORITHM

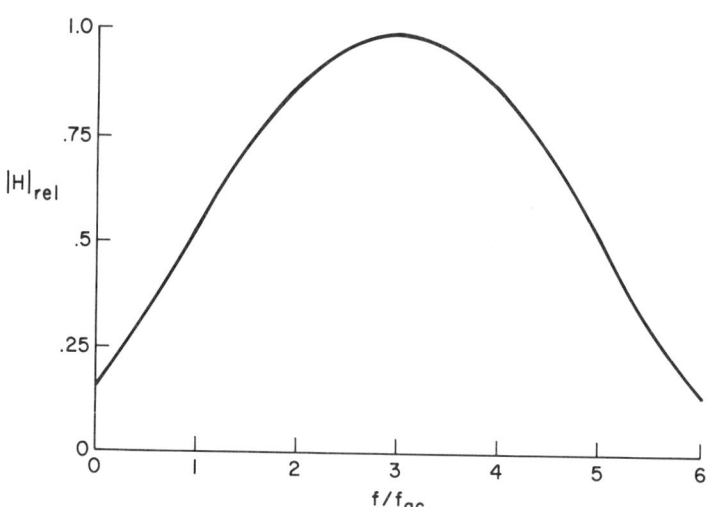

5. FIRST AND SECOND DERIVATIVE ALGORITHM

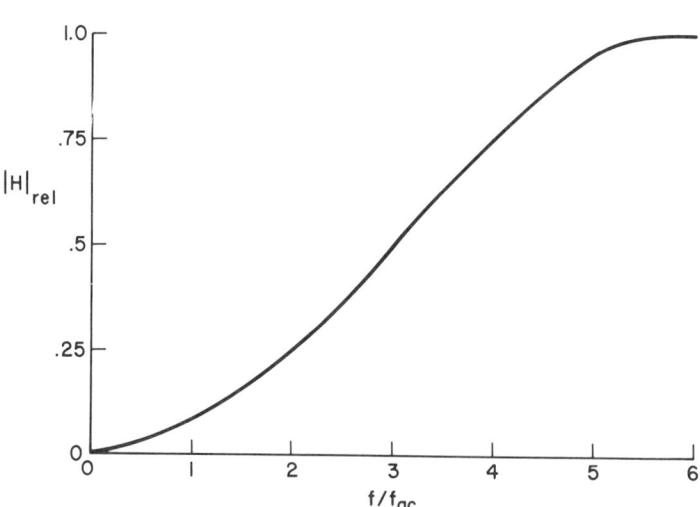

6. SOLUTION OF DIFFERENTIAL EQUATION

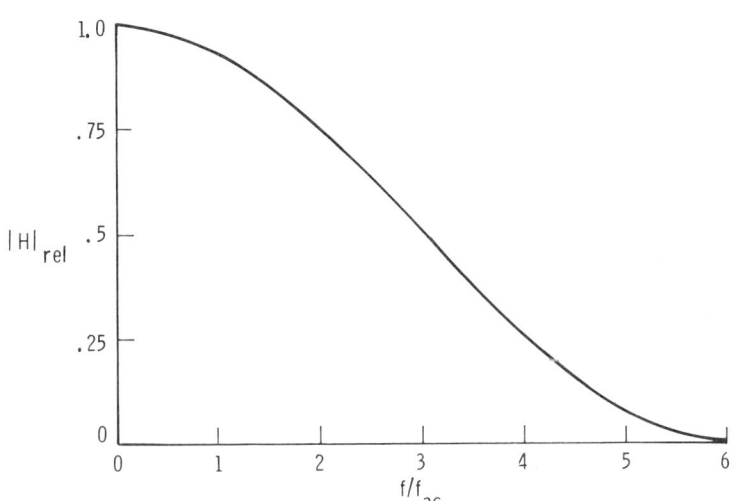

7. SOLUTION OF DIFFERENTIAL EQUATION WITH SELECTED LIMITS OF INTEGRATION

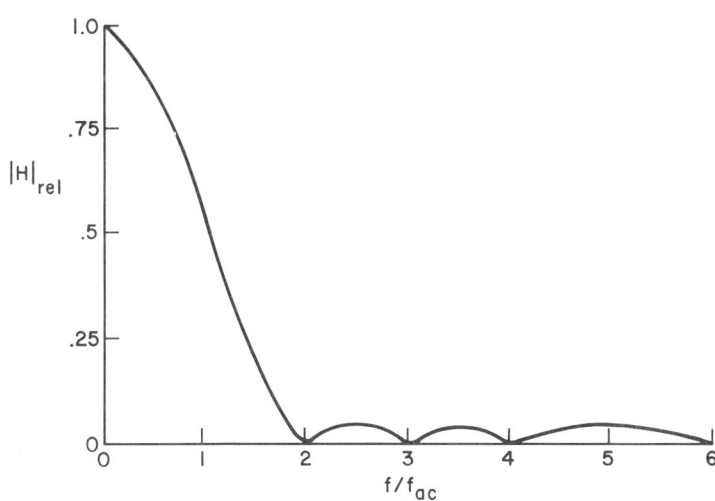

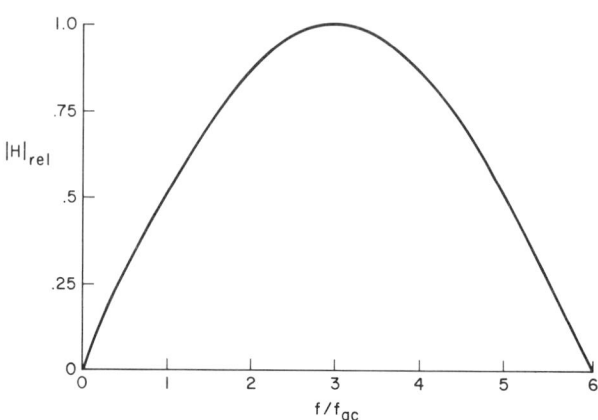
8. SINUSOIDAL CURVE FIT

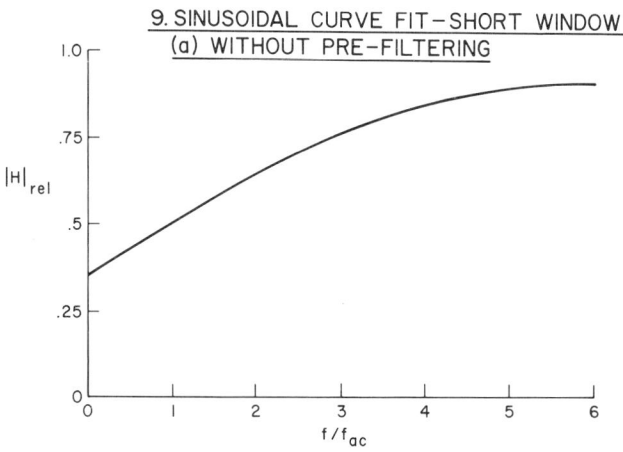
9. SINUSOIDAL CURVE FIT – SHORT WINDOW
(a) WITHOUT PRE-FILTERING

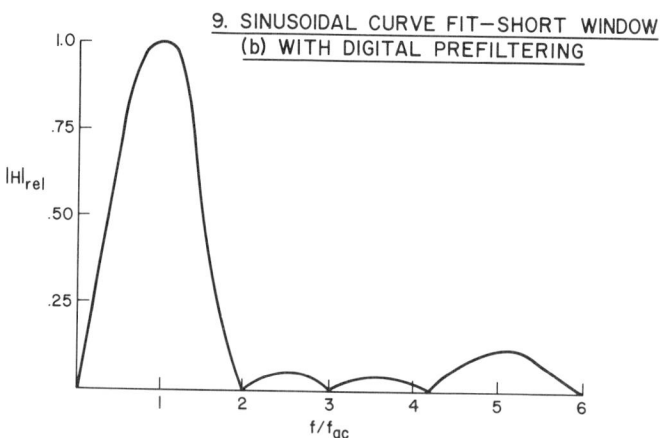
9. SINUSOIDAL CURVE FIT – SHORT WINDOW
(b) WITH DIGITAL PREFILTERING

10. LEAST-SQUARES FIT

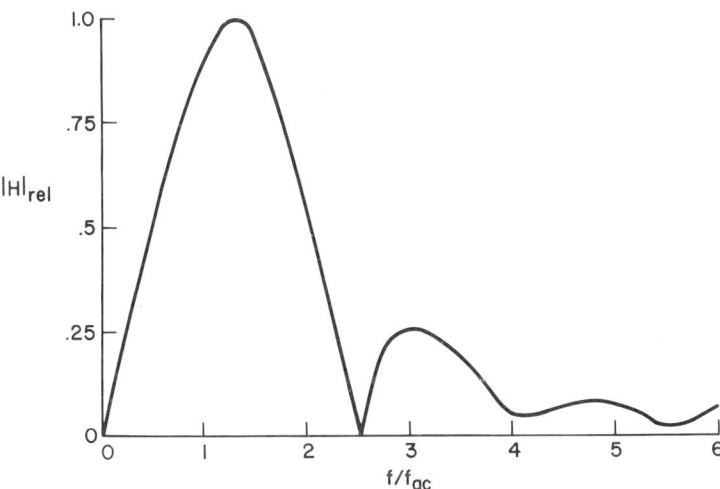

Fig. 2. Frequency-Response Plots of the Algorithms Tested

1. Fourier analysis with one-cycle window (Ramamoorty and others, [7]) - the incoming ac data samples for one cycle are correlated with the stored samples of reference fundamental sine and cosine waves to extract the complex value of the fundamental component in rectangular form. The general expressions for the sine and cosine components of voltage at a sample point k are

$$V_s = \frac{1}{N}\left[2 \sum_{\ell=1}^{N-1} v_{k-N+\ell} \sin\left(\frac{2\pi}{N}\ell\right)\right]$$

$$V_c = \frac{1}{N}\left[v_{k-N} + v_k + 2 \sum_{\ell=1}^{N-1} v_{k-N+\ell} \cos\left(\frac{2\pi}{N}\ell\right)\right]$$

where the v_i are the voltage samples and N is the number of samples taken per fundamental cycle. Similar expressions are evaluated for current components I_S and I_C; the four results can be used to generate the phasor impedance value in polar or rectangular form. In polar form,

$$|Z| = \sqrt{\frac{V_s^2 + V_c^2}{I_s^2 + I_c^2}}$$

$$\theta = \arctan(I_s/I_c) - \arctan(V_s/V_c).$$

Implicit in Fourier analysis is drastic filtering of the data; the output responds slowly, smoothly, and accurately to badly distorted fault waveforms.

The frequency response plot shows the response peak at the power system frequency, with nulls at dc and at each harmonic. High frequencies above the second harmonic are all well attenuated.

2. Fourier analysis with short data window (Phadke, Hlibka, and Ibtahim [6]) - the basis for this method is the same as (1) above, but the data window is shortened to 1/2 cycle plus one sample for faster response:

$$V_s = \frac{4}{N} \sum_{\ell=1}^{N/2} v_{k-(N/2)+\ell} \sin\left(\frac{2\pi}{N}\ell\right)$$

$$V_c = \frac{4}{N} \sum_{\ell=1}^{N/2} v_{k-(N/2)+\ell} \cos\left(\frac{2\pi}{N}\ell\right).$$

Of course, accuracy of results is more seriously affected by off-normal frequency components; dc offsets present a particular problem. The authors of the algorithm remedy the latter weakness by assuming that the fault waveform contains a dc offset of unknown magnitude but with time constant determined by the known X/R ratio of the line. The suggested implementation effectively finds the magnitude of the offset and subtracts it from the fault waveform prior to the Fourier analysis itself.

The response plot, when compared to the full cycle Fourier analysis, shows reduced effectiveness in coping with dc and even harmonics as expected. This is overcome, in part, by the dc offset correction.

3. Walsh Analysis (Horton [10]) - the Walsh algorithm is closely related to the Fourier analysis of (1) above. The orthogonal functions which are correlated with the fault waveform, however, are not fundamental sine and cosine waves, but odd and even square waves. Real-time computation is simplified since the reference square waves assume values of ±1 only. However, some of this benefit is mitigated by the need to extract several harmonics of the square-wave components

along with the fundamental so that the desired fundamental sinusoidal component can be reconstructed.

As for Fourier analysis, the filtering of the source data is drastic, and the resultant output is thus heavily damped. Accuracy is good even for distorted wave forms.

Note that the frequency response plot is given for a sampling rate of eight per cycle, rather than 12 as in the other cases; the Nyquist limit coincides with the fourth harmonic. This is because the Walsh method is convenient to implement only for a number of samples which is an integral power of two. Although this plot is not directly comparable with the Fourier plot (1) above, the two algorithms, in fact, behave almost identically if implemented for the same sampling rate.

4. Sample and Derivative Calculation (Mann and Morrison [2]) - this algorithm was the first proposed for the characterization of a sinusoidal waveform from asynchronous data samples. If the equations for the instantaneous value of the fundamental sinusoid (v = V sin(wt +)) is differentiated, the result and the original equation can be solved to yield the peak magnitude or phase in terms of the value and slope of the waveform at an arbitrary sampling instant:

$$|V| = \sqrt{(v_k)^2 + (v_k')^2}$$

$$\theta_v = \arctan\left[\frac{v_k}{v_k'}\right].$$

The slope is approximated by the difference between two samples straddling the central sample point:

$$v_k' = \frac{1}{h\omega}\left[v_{k+1} - v_{k-1}\right].$$

where h is the sampling interval in seconds and ω is the fundamental angular frequency. With a window of only three samples, the algorithm responds quickly to sudden changes in the waveform. Dc offsets, however, upset the expected relationship between the sample value and its first difference by changing only the former. Similarly, harmonics and other high frequencies can produce large errors in the first difference without much change of the central value. In either case, the calculated peak value and phase are inaccurate. The authors of the algorithm attenuate offsets in the current with an R-L shunt network on the ct secondary whose X/R ratio matches that of the line (analog mimic burden). High frequencies are attenuated with a digital smoothing or filtering routine.

The investigators tested the algorithm both with and without a digital filter whose characteristic is identical to the analog mimic burden.

The response plot shows that the peak occurs at the third harmonic, with no nulls at all. High frequency contamination can thus cause severe errors.

5. First and Second Derivative Calculation (Gilcrest, Rockefeller, and Udren [4]) - if the same derivation as in (4) above is applied to the first and second derivatives of the equation for the sinusoid rather than to the equation itself and the first derivative, the result is:

$$V = (v_k') + (v_k'')$$

$$v = -\arctan \frac{v_k''}{v_k'}$$

where

$$v_k'' = \frac{1}{(h)} (v_{k+1} - 2v_k + v_{k-1})$$

and

$$v_k' = \frac{1}{h} v_{k+1} - v_{k-1} .$$

The key benefit is that a constant dc offset now has no effect since it drops out from all difference expressions. While exponentially decaying offsets have non-zero frequency components which do cause error, the effect is minimal. No mimic burden or related technique is needed. Response is as fast as in (4).

The disadvantage is that above-normal frequencies produce errors in all of the first and second differences, causing

DEVELOPMENT AND SELECTION OF ALGORITHMS

even larger errors than in (4) above. In the frequency-response plot, compare the response at the power system frequency ($f/f_{ac} = 1$) to that at the Nyquist limit ($f/f_{ac} = 6$). Unsuppressed harmonics can destroy the usefulness of the results.

6. Solution of Differential equation (McInnes and Morrison [12]) - the transmission line being protected is modeled as a series R-L circuit resulting in the following equation:

$$v = R_{eff} i + L_{eff} \frac{di}{dt} .$$

Solution for the R and L parameters is accomplished by integration over two successive time periods and solution of the resulting simultaneous linear equations. Integrations are performed using the trapezoidal rule.

This algorithm has the advantage of recognizing dc offsets as valid components of fault currents rather than processing them as unaccounted-for. The assumption on which it is based is valid for lines that are not extremely long, permitting shunt capacitance to be neglected.

The differential-equation method was implemented for two different data windows at each sampling rate - 3- or 5- sample windows for sampling rates of eight or 12 per cycle, and 3- or 17- sample windows at 32 samples per cycle.

The plot shown is for a data window of five samples. High-frequency response falls off to a null at the Nyquist limit. The large response to dc and low frequencies does not reflect unfavorably on the ability of the algorithm to deal with the exponentially decaying dc offset whose occurrence is consistent with the line model.

7. Solution of Differential equation with Selected Limits (Ranjbar and Cory [13]) - incoming data is assumed to be a solution set of a differential equation for a series R-L model of a transmission line as in (6). The R and L parameters are calculated from the equations below by selecting overlapping limits of integration chosen to eliminate specific low-order harmonics and their multiples.

For example, to eliminate harmonics, K, M, and N the authors give

$$L\left(\sum \int di_y\right) + R\left(\sum \int i_x dt\right) = \sum \int v dt$$

$$\sum \int v dt = \int_0^{\frac{2\pi}{K}} v dt + \int_{\frac{\pi}{M}}^{\frac{2\pi}{K}+\frac{\pi}{M}} v dt$$

$$+ \int_{\frac{\pi}{N}}^{\frac{2\pi}{K}+\frac{\pi}{N}} v dt + \int_{\frac{\pi}{M}+\frac{\pi}{N}}^{\frac{2\pi}{K}+\frac{\pi}{M}+\frac{\pi}{N}} v dt .$$

The quantities $\sum \int di_y$ and $\sum \int i_x dt$ are calculated similarly. Ability to properly process dc offsets is retained and errors due to low order harmonics are minimized. For complete suppression, the sampling rate must be a multiple of the harmonic order to be eliminated. The frequency-response form plot shows the improved suppression of high frequencies.

8. Sinusoidal Curve Fit (Gilbert and Shovlin [14]) - incoming data is used to directly calculate apparent resistance and reactance to the fault. The algorithm is based on fitting data to fundamental sinusoidal quantities (voltage and current) using three consecutive samples. Equations for X and R are

$$X = \frac{v_{n-1} i_n - v_n i_{n-1}}{(i_{n-1}^2 - i_{n-2} i_n) \csc \Delta}$$

$$R = \frac{2 v_{n-1} i_{n-1} - v_n i_{n-1} - v_{n-2} i_n}{2(i_{n-1}^2 - i_{n-2} i_n)}$$

where Δ is the angular displacement between samples.

The equations respond rapidly to changes in the incoming data, but are highly susceptible to high frequency transients. Note the sensitivity to the third harmonic in the transfer functions.

9. Sinusoidal Curve Fit with short data window (Makino and Miki [15]) - voltage and current are assumed to be pure

sine waves at the funadmental system frequency. Incoming data is used to calculate peak values of voltage and current, as well as power flow, using the following equations:

$$|V|^2 = \frac{v_1^2 + v_2^2 - 2v_1v_2 \cos \omega t}{(\sin \omega t)^2}$$

$$|I|^2 = \frac{i_1^2 + i_2^2 - 2i_1i_2 \cos \omega t}{(\sin \omega t)^2}$$

$$VI \cos \theta = \frac{v_1i_1 + v_2i_2 - (v_2i_1 + v_1i_2) \cos \omega t}{(\sin \omega t)^2}.$$

These equations will exhibit errors in proportion to individual sample errors due to the two-sample window. Digital filters are proposed by Makino and Miki for preprocessing of data to attentuate dc offsets and harmonics of third and higher order. Tests were conducted both with and without these filters. Two plots are given, showing the response of the basic algorithm, and the drastic reduction in off-normal response provided by the filters. Such filtering routines can, of course, be applied to any algorithm with a considerable speed penalty.

This algorithm can be shown to be mathematically equivalent to the sinusoidal curve fit algorithm suggested by Gilbert and Shovlin in (8) above.

10. Least-Squares Fit (Luckett, et. al. [16]) - data samples are used to obtain a least-squares fit to an assumed waveform containing sinusoidal and exponentially decaying components. The assumed waveform is:

$$K_1 e^{-\lambda t} + \sum_{m=1}^{N} (K_{2m} \sin m\omega t + K_{2m+1} \cos m\omega t)$$

where the solution for the K values is found by a least-squares fit of the form which minimizes the mean-square error between the assumed and actual waveforms:

$$K_r = \int_0^T I\,fr(t)\,dt \qquad \text{for } r = 1, 2, \ldots, 2N + 1.$$

Few additional details regarding the subject algorithm are presented in reference 16. Thus, there are many least-squares fits possible, including a Fourier Transform solution (which can also be shown to be a least-squares fit).

The transfer-function plot describes the response with a 10-sample window and 12 samples taken per cycle. It shows a peak near, but not exactly at, the fundamental frequency, with lesser response to high frequencies. Nulls can be introduced at harmonic frequencies by including harmonic terms in the assumed waveform expression above, but this severely complicates the computations.

In evaluating the least-squares fit, two windows were tried at each sampling rate. At eight samples per cycle, 5- and 7-sample windows were used. For 12 samples per cycle, the windows were seven and 10 samples, and at 32 samples per cycle, 17 and 25 samples were included in the window.

Of the ten algorithms just characterized, two (sample-and-derivative and short-window sinusoidal fit) were tested with and without prefiltering routines; two others (differential-equation and least-squares fit) were tested with multiple data windows. Thus, a total of 14 variations were tested at each sampling rate.

VII. TESTING METHODS

The prime source of test data for this evaluation is an analytic fault program written in BASIC and executed on the Univac Data Processing System at the Westinghouse Research Laboratories. The program accepts as inputs specifications of pre- and post-fault parameters of voltage and current in such a way as to produce fault waveforms which are consistent with an assumed R-L model of the line. Fault inception angles in each case are specified to produce full, half, or zero offset in the currents, as desired. Other transient distortion components can be specified and introduced in a strictly controlled and artificial manner. These other components include low-order harmonics (no decrement), a 600 Hz signal representing the fundamental component of a series of wave fronts reflected between the fault and line terminals, and subnormal frequency components which might be produced by faults on series-compensated lines.

Instead of using these analytically-generated and somewhat artificial waveforms, the authors could have chosen to use, as many other investigators have, a digital-computer based model of an EHV line which includes an array of lumped inductive, resistive, and shunt capacitive elements. Such a model produces fault waveforms which are totally realistic to the extent that the model represents the performance of a real, distributed line. The former approach was selected because:

1. The precise distortion content is known and completely controllable. This allowed the authors to see the response of the algorithm to specific transients introduced one at a time. This more clearly pinpoints specific strengths, weaknesses, and levels of tolerance for each algorithm and it suggests possible fixes for each, such as specific types of filtering.

2. None of the algorithms assumed a specific relationship between the fundamental quantity and high frequency transient components. Therefore, the validity of results is not lessened by the artificial relationship between these transients and the fundamentals.

3. The authors investigated at some length the relationship between published fault waveforms produced by digital computer line models and those on oscillograms from actual EHV lines. Even considering the filtering effects of iron-core ct's and the oscillographs themselves, the real-system waveforms contained only a small fraction of the harmonics and other high frequencies produced by some models. Checking the literature revealed that remarkably little is known about the spectral content of fault waveforms from actual transmission lines. The authors concluded that, while worst-case harmonics must be assumed to be larger than those seen on typical oscillograms, the model line waveform represents an overly severe test which, if used as a key selection criterion, would lead to the choice of a slower-than-necessary algorithm.

After the evaluation described here, the authors did test selected algorithms using data from a 600-volt analog model power system, as well as samples of faults from an actual 230 kV line produced by the experimental system of [4]. The results of these tests support the conclusions reached using the analytic data. The authors reserve the detailed results of these tests for a future work.

The data samples generated by the fault program are stored in the Univac system data files for use by algorithm-testing programs. Sample values and a simple reference plot are printed out as shown in the examples of Figure 3. 3 (a) shows a 60 Hz waveform, samples 12 times per cycle for three cycles, with fault inception one cycle from the start. The current exhibits dc offset but is otherwise pure. In 3 (b), substantial third and fifth harmonics are present in the voltage and current after the fault. Printed at the top of each plot, but not shown in Figure 3, is a list of fault parameters specified for that run.

VIII. SPECIFIC TESTS USED

The 14 algorithms were each subjected to the following faults. Each fault is preceded by heavy load flow into the line except as noted; in all cases but one, the line angle was assumed to be 85°.

1. Inception angle chosen for full current offset.
2. Half offset.
3. No offset.
4. 40% third and 20% fifth harmonic (prefault base) on voltage and current.
5. 15% third and 7.5% fifth harmonic.
6. 5% third and 2.5% fifth harmonic.
7. 1.66% third and 0.83% fifth harmonic.
8. 10% 600 Hz on voltage and current (prefault base).
9. Reverse prefault load flow direction.
10. Low load before fault; full, half, and no offset.
11. 75° line angle.
12. Fault with resistance to produce 45° fault angle; full, half, and no offset.
13. Transformer inrush, generated by a separate program which modeled the behavior of a transformer core to produce a simulated inrush current waveform for an unfavorable combination of residual flux and closing angle.
14. Extended prefault interval (for checking steady-state accuracy).
15. Fault with 20 Hz subnormal components in the voltage and current similar to those produced by series-compensated lines.

DEVELOPMENT AND SELECTION OF ALGORITHMS 105

For each of these faults, the program produced data at 32, 16, 12, eight, and four samples per cycle, although sampling rates and algorithms were selectively paired. In each case, two cycles of fault samples follow one cycle of pre-fault data, except for the extended pre-fault case. Finally, the authors tested the effects of simulating A/D converters of decreasing resolution by quantizing the fault samples. These results were evaluated subjectively and were not included in the matrix evaluation which follows.

IX. RAW TEST RESULTS

A large BASIC program, implemented on the previously-mentioned Univac Data Processing System, applied each of the algorithms in turn to the fault data base. The resultant printout for the Fourier algorithm with fully offset current at fault inception and a sampling rate of 32 per cycle is shown in Figure 4. No other transients are included in this test fault. The output of each algorithm is expressed as a complex impedance in rectangular form (R + jX); the calculated values of R and X are shown as separate functions of time. The plot begins with the algorithm results from the last pre-fault sample, and shows the change in calculated impedance as the algorithm data window moves into the fault. The plotting continues for two cycles after the fault inception point.

Although the resolution of the plot is limited by the column width of the line printer, the exact calculated values are printed on the left side for reference when needed. A key data output is the statistical package shown just above the plot. The means and standard deviations for R and X_L are given, with the data included in the statistical calculation beginning one-half or one cycle after fault inception. Results are expressed both as a percentage of the expected post-fault values, and as a percentage of an assumed transmission line whose electrical length is three times that from the relaying point to the fault (the fault location, and thus the length of the assumed line, are held constant for these tests). The expected post-fault values are defined here as those which would be obtained from the algorithm if the fault data were free of all transients and quantization. When such pure data is used, all algorithms yield the same expected values if the data window has moved entirely into the post-fault interval. Therefore, the statistical errors result from response to transients, quantization of data, and delay in response of the algorithm after fault inception because of a long data window.

```
OUTFILE:    OUTDAT100

TIME-MS  VOLTS   AMPS
 -16.7     9.4   -2.9
 -15.3    56.9     .7
 -13.9    89.2    4.1
 -12.5    97.5    6.4
 -11.1    79.8    7.0
  -9.7    40.6    5.8
  -8.3    -9.4    2.9
  -6.9   -56.9    -.7
  -5.6   -89.2   -4.1
  -4.2   -97.5   -6.4
  -2.8   -79.8   -7.0
  -1.4   -40.6   -5.8
   0.0     3.1   -2.9
   1.4    19.0    3.5
   2.8    29.7   17.2
   4.2    32.5   34.2
   5.6    26.6   49.5
   6.9    13.5   58.6
   8.3    -3.1   58.9
   9.7   -19.0   50.0
  11.1   -29.7   33.9
  12.5   -32.5   14.6
  13.9   -26.6   -2.8
  15.3   -13.5  -14.1
  16.7     3.1  -16.4
  18.1    19.0   -9.3
  19.4    29.7    5.0
  20.8    32.5   22.4
  22.2    26.6   38.3
  23.6    13.5   47.9
  25.0    -3.1   48.7
  26.4   -19.0   40.2
  27.8   -29.7   24.5
  29.2   -32.5    5.7
  30.6   -26.6  -11.3
  31.9   -13.5  -22.2
  33.3     3.1  -24.2
```

Fig. 3. Outputs of Analytic Fault Program

DEVELOPMENT AND SELECTION OF ALGORITHMS

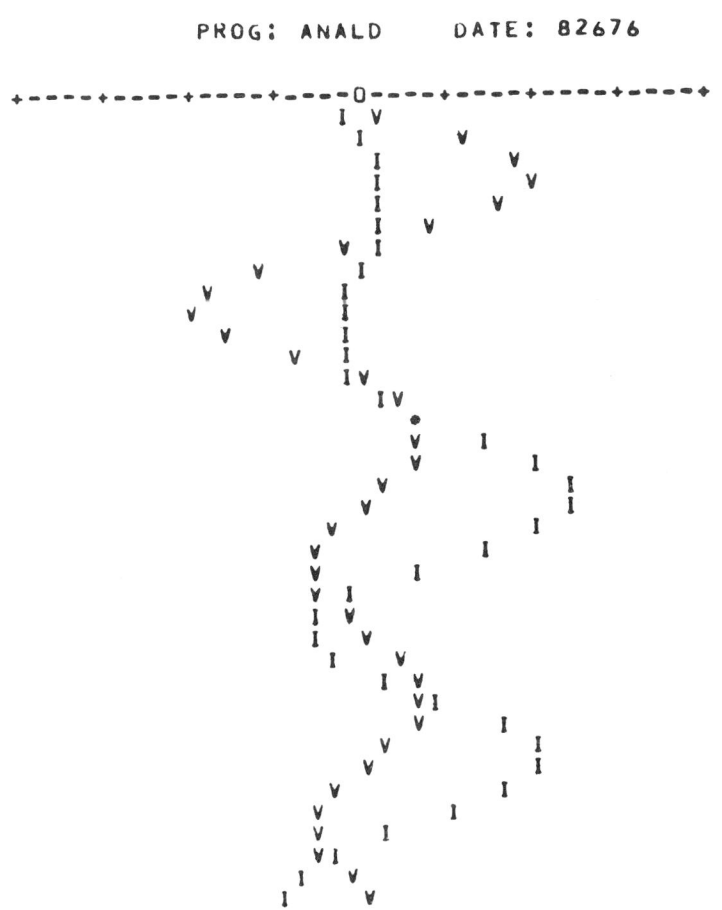

(a) Fundamental and Dc Offset

```
OUTFILE:    OUTDAT104

TIME-MS  VOLTS   AMPS
 -16.7    97.5    6.4
 -15.3    79.8    7.0
 -13.9    40.6    5.8
 -12.5    -9.4    2.9
 -11.1   -56.9    -.7
  -9.7   -89.2   -4.1
  -8.3   -97.5   -6.4
  -6.9   -79.8   -7.0
  -5.6   -40.6   -5.8
  -4.2     9.4   -2.9
  -2.8    56.9     .7
  -1.4    89.2    4.1
   0.0    32.5    6.4
   1.4    45.0   24.3
   2.8     7.2   32.9
   4.2   -10.5   34.2
   5.6   -25.3   26.4
   6.9   -11.3   13.1
   8.3   -32.5   -6.4
   9.7   -45.0  -24.3
  11.1    -7.2  -32.9
  12.5    10.5  -34.2
  13.9    25.3  -26.4
  15.3    11.3  -13.1
  16.7    32.5    6.4
  18.1    45.0   24.3
  19.4     7.2   32.9
  20.8   -10.5   34.2
  22.2   -25.3   26.4
  23.6   -11.3   13.1
  25.0   -32.5   -6.4
  26.4   -45.0  -24.3
  27.8    -7.2  -32.9
  29.2    10.5  -34.2
  30.6    25.3  -26.4
  31.9    11.3  -13.1
  33.3    32.5    6.4
```

Fig. 3. Outputs of Analytic Fault Program

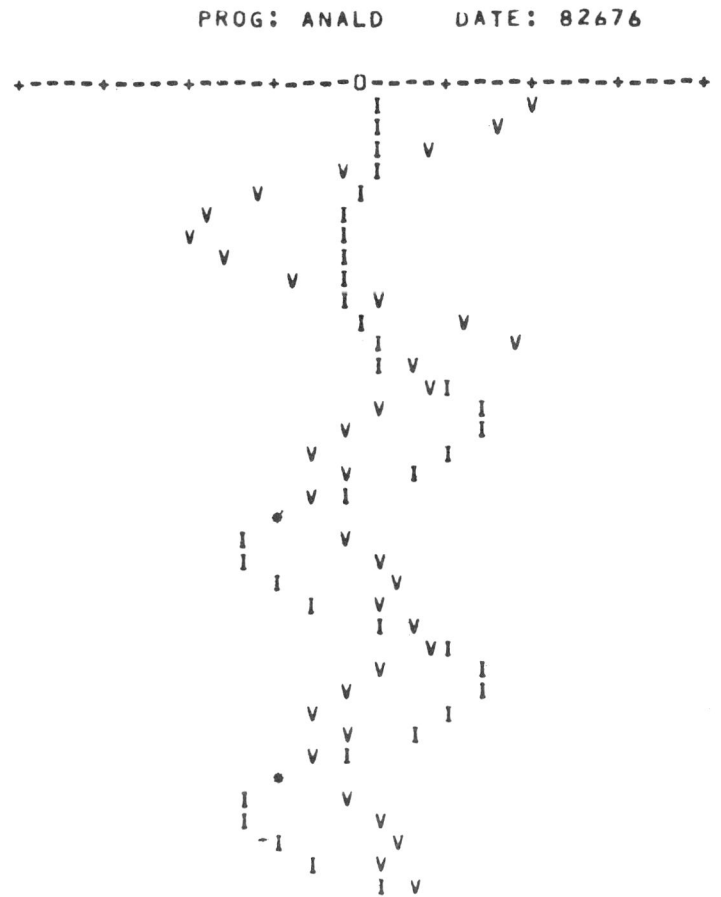

(b) Fundamental with Third and Fifth Harmonics

DATA FILE: OUTDATOOO   PROG: BALA32   DATE: 72876
STATISTICS TAKEN OVER ONE POST FAULT CYCLE
BEGINNING AT 1 CYCLE AFTER FAULT INCEPTION
            MEAN(X10)   STD DEV(X10)   ERROR OF
XL   9.130   .599   -.790
R   .876   .670   8.850
BEGINNING AT 1/2 CYCLE AFTER FAULT INCEPTION
            MEAN(X10)   STD DEV(X10)   ERROR OF
XL   10.629   2.494   15.499
R   5.128   6.047   536.923

                                                    XL=•
PT   XL(X10)   R(X10)   -10   -8   -6   -4
                                1.........1.........1.........1....

0   68.558   122.066
1   69.032   120.116
2   67.939   120.451
3   65.036   121.129
4   58.998   121.865
5   48.872   121.088
6   35.359   116.208
7   21.631   105.532
8   11.501   90.429
9   6.423   74.366
10   5.287   60.064
11   6.286   48.473
12   8.083   39.437
13   9.957   32.452
14   11.592   27.020
15   12.885   22.749
16   13.830   19.353
17   14.458   16.623
18   14.810   14.409
19   14.926   12.600
20   14.840   11.111
21   14.583   9.871
22   14.179   8.823
23   13.652   7.913
24   13.027   7.093
25   12.338   6.321
26   11.626   5.562
27   10.943   4.800
28   10.341   4.045
29   9.859   3.335
30   9.507   2.725
31   9.259   2.266
32   9.064   1.990
33   8.880   1.848
34   8.710   1.733
35   8.571   1.598
36   8.463   1.450
37   8.384   1.293
38   8.333   1.131
39   8.309   .969
40   8.312   .809
41   8.338   .654
42   8.389   .506
43   8.463   .370
44   8.558   .247
45   8.673   .141
46   8.806   .056
47   8.954   -.005
48   9.114   -.038
49   9.280   -.040
50   9.446   -.008
51   9.605   .058
52   9.748   .158
53   9.868   .288
54   9.956   .442
55   10.006   .612
56   10.017   .787
57   9.988   .957
58   9.922   1.112
59   9.826   1.244
60   9.707   1.349
61   9.574   1.422
62   9.435   1.463
63   9.297   1.475
64   9.165   1.460

Fig. 4. Example Test Result - Full-Cycle Fourier Algorithm,

DEVELOPMENT AND SELECTION OF ALGORITHMS

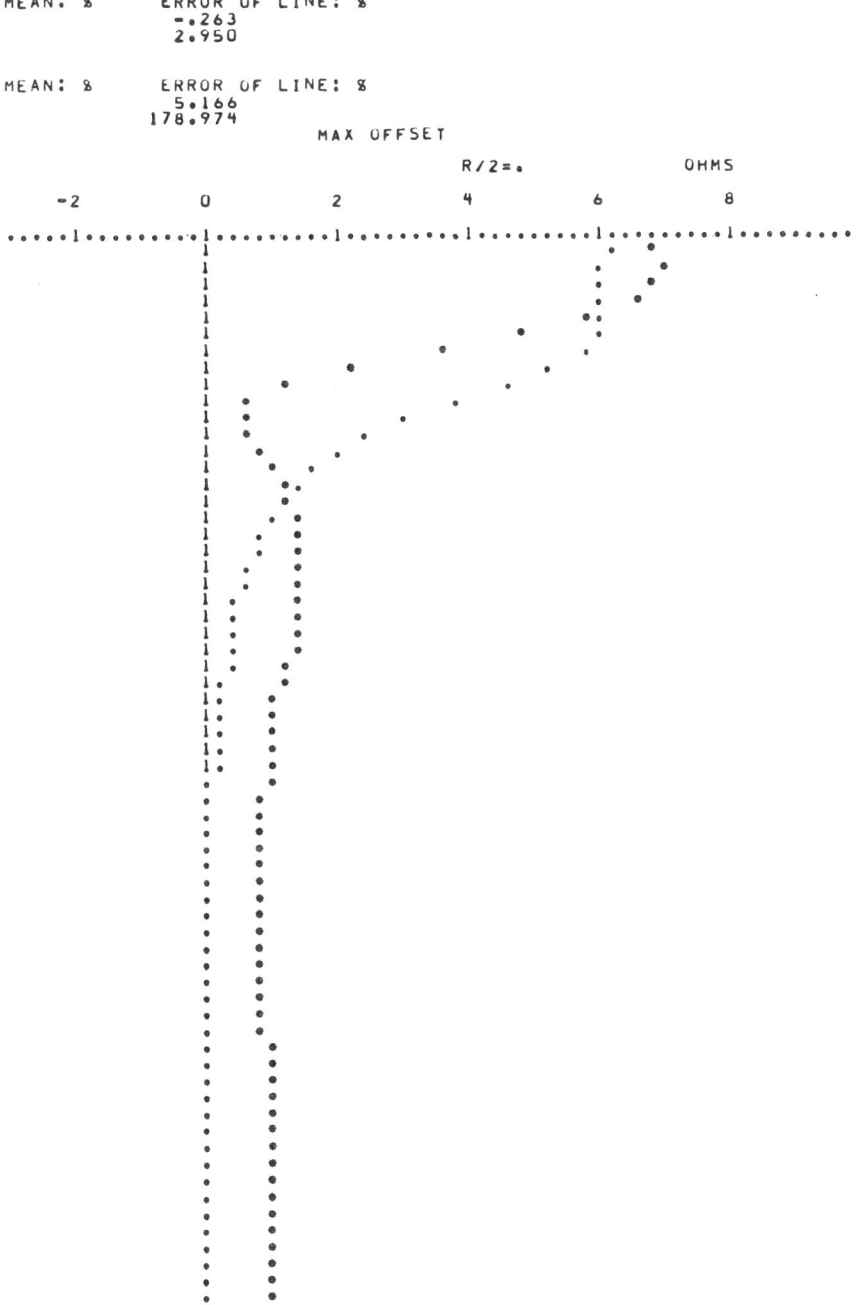

32 Samples Per Cycle, Full Offset in Test Current

The authors recognize that care must be used in interpreting mean and standard deviation results when the data analyzed is not known to be normally distributed. Still, the statistical results were extremely useful in evaluating test cases containing severe harmonics. In such cases, the plot often showed calculated impedances scattered over the entire page. Thus, visual inspection was not useful. The statistical package, though, could show whether the average of these values over time was still reasonably accurate and useful. When the mean is reasonable, the standard deviation provides a fair measure of the relative scatter of the results for a particular algorithm. The statistics guided the authors in the decision-making process described below, as well as in the design of auxiliary filters.

Figures 5 through 8 show results for other algorithms and faults, as simple examples of the performance tradeoffs explained earlier. In Figure 5, the Mann-Morrison algorithm with a window of only three samples and a digital version of the dc offset mimic-burden filter processes the same fault as was used in Figure 4. Not surprisingly, the Mann-Morrison method produces almost perfectly accurate results starting four samples after fault inception, while in the Fourier analysis case the results do not stabilize near the expected values until the full one cycle (32-sample) data window moves into the post-fault interval. Using this fault alone for comparison, the slow-acting Fourier algorithm is without merit. But now, study the results in Figures 6 and 7. Here, the offset test has been replaced by the 15% third and 7.5% fifth harmonic test. The Fourier algorithm, demonstrating its filtering ability, still converges slowly and steadily on the expected values, yielding very low errors after one cycle. The broadbanded Mann-Morrison algorithm, on the other hand, shows the harmonic content in its output and provides only marginally useful results.

Finally, Figure 8 shows the response of the Prodar 70 algorithm (whose frequency-response plot shows a tendency to severely accentuate high frequencies) to a fault with heavy harmonic content. Obviously, such flypaper-like results are useless.

In virtually all cases, as in the examples just given, the performance traits predicted by analysis were confirmed by the fault tests. Nonetheless, several unanticipated idiosyncrasies of some algorithms were found during the testing.

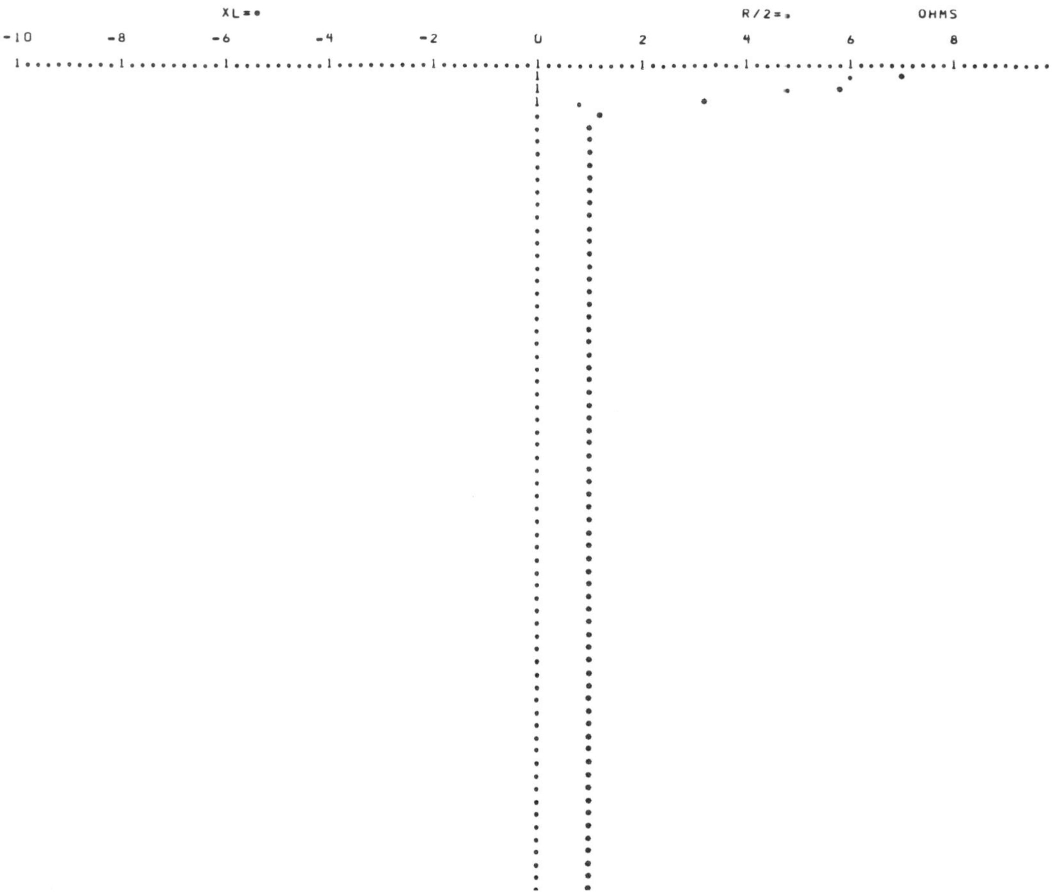

Fig. 5. Test Result for Mann-Morrison Algorithm with Mimic Filter, Full Offset in Test Current

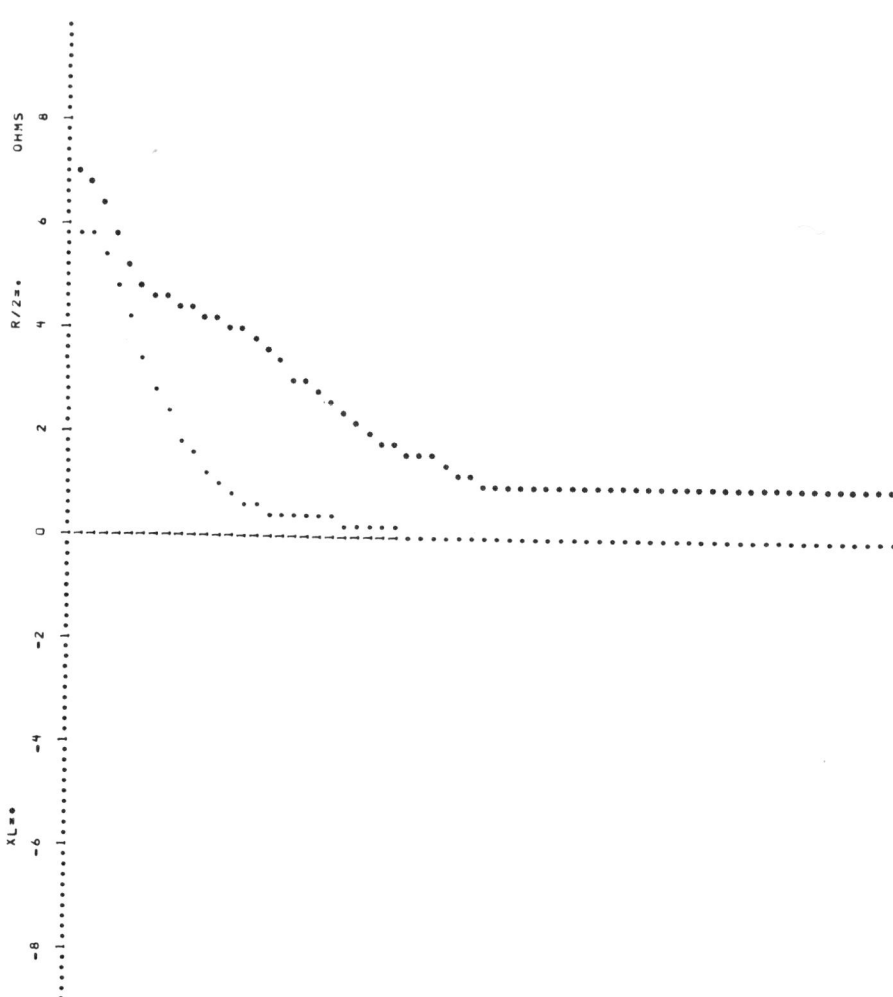

Fig. 6. Test Result for Fourier Algorithm, Third

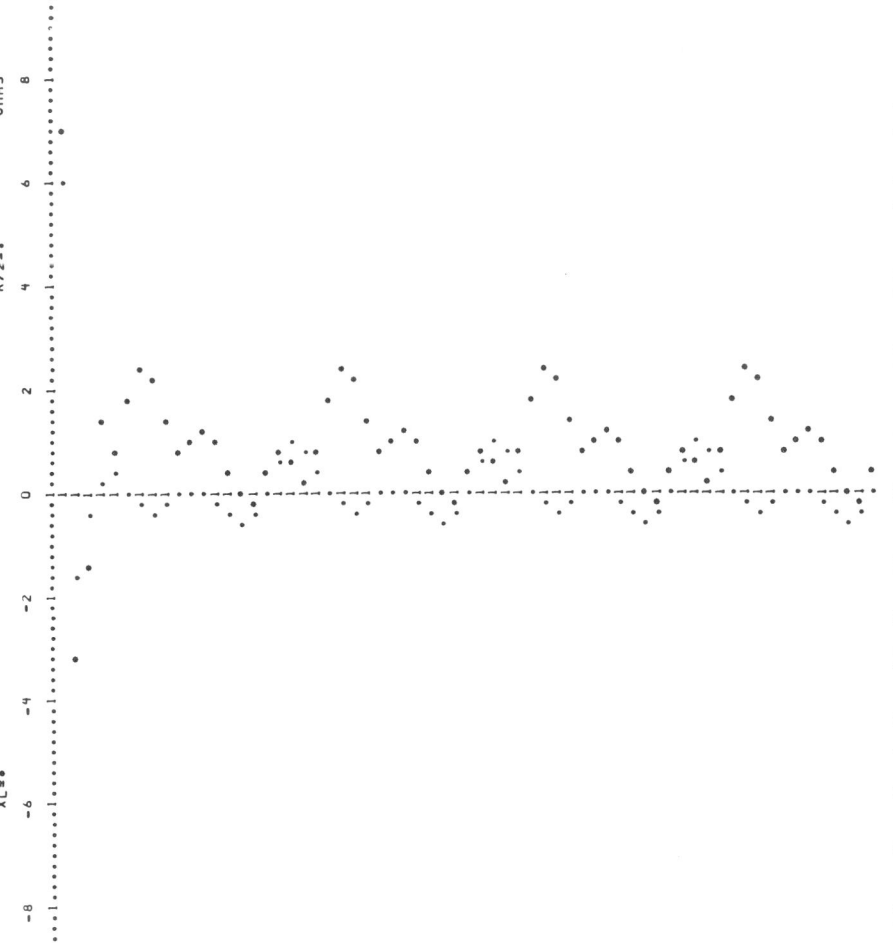

Fig. 7. Test Result for Mann-Morrison Algoritm with Mimic Filter, Third and Fifth Harmonics in Test Quantities

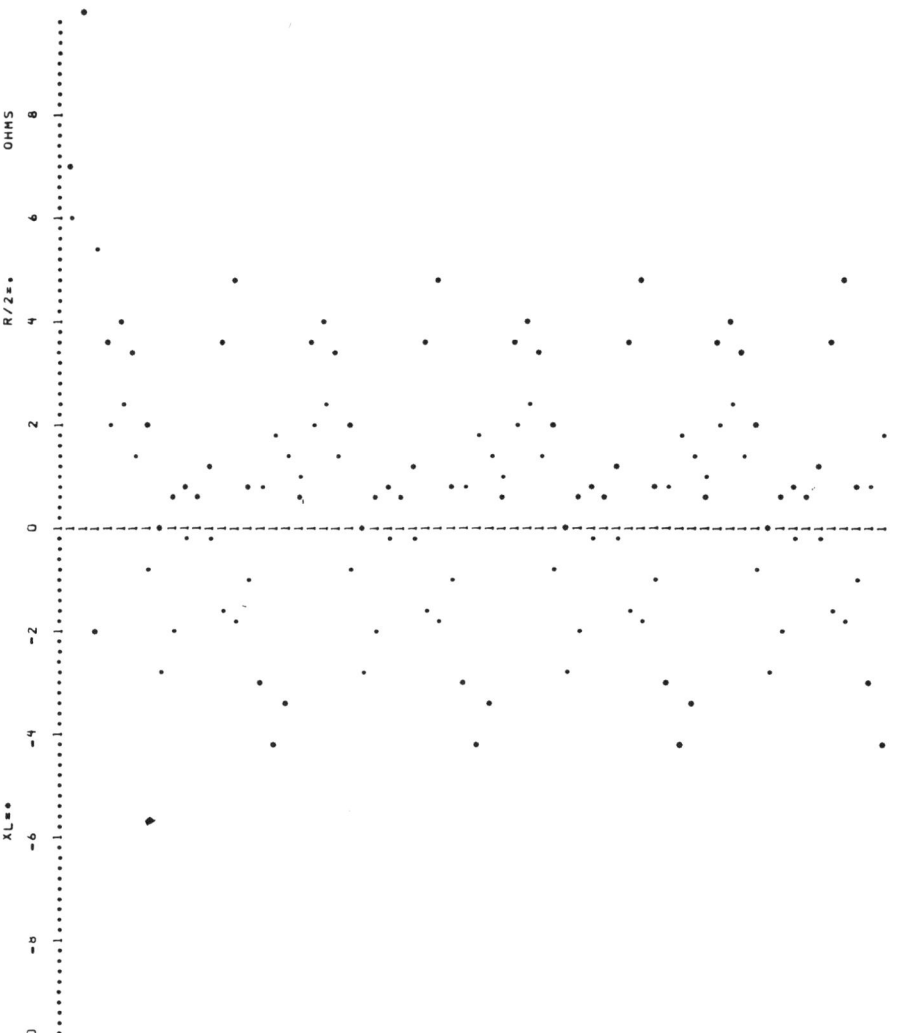

DEVELOPMENT AND SELECTION OF ALGORITHMS

With the number of faults, algorithms, and sampling rates used, nearly a thousand test runs like those of Figures 4 through 8 were produced. The authors quickly discovered that evaluation of the results became entirely subjective - an individual could not retain key observations made at the beginning of the stack as he progressed toward the end. What was needed was a technique of setting fixed criteria, based on the investigators' special requirements, and then objectively and uniformly applying them to all of the algorithm tests. The following describes such a procedure.

Stochastic Ranking and Matrices

To evaluate the data uniformly and fairly, the authors have applied a process similar to that termed Decision Analysis by Kepner and Tregoe [19]. The heart of this technique is formation of a matrix having rows of alternatives and columns of criteria against which the alternatives are measured. Criteria are weighted in proportion to their importance and each alternative is ranked according to how well it meets each criterion. The products of criterion weights and assigned rank are summed to form total scores. The highest score best meets the composite requirements defined by the individual criteria.

A series of six matrices was used to process the statistical data from a selected group of tests into rankings based on total scores. The matrices were identified as:

1. 32 samples per cycle; results beginning one cycle after fault inception.
2. 32 samples per cycle; results beginning one-half cycle after fault inception.
3. Same as (1) for 12 samples per cycle.
4. Same as (2) for 12 samples per cycle.
5. Same as (1) for eight samples per cycle.
6. Same as (2) for eight samples per cycle.

The format of one matrix is shown in Figure 9.

Each has 14 major rows corresponding to 14 algorithms considered: statistical results for X_L and R output of each of the algorithms are entered separately. There are nine columns which correspond to nine test conditions considered most important: maximum offset, low level pre-fault conditions, all but the most severe harmonic tests, 600 Hz tests, steady-state accuracy (extended prefault) and reverse prefault load, and finally, a speed column which listed the number of samples for R and X_L to fall within a specified band of the expected values of each. The latter was found from visual assessment of the plot and output data listing. Each of the nine columns is assigned a weighting factor according to the importance it holds for the investigators.

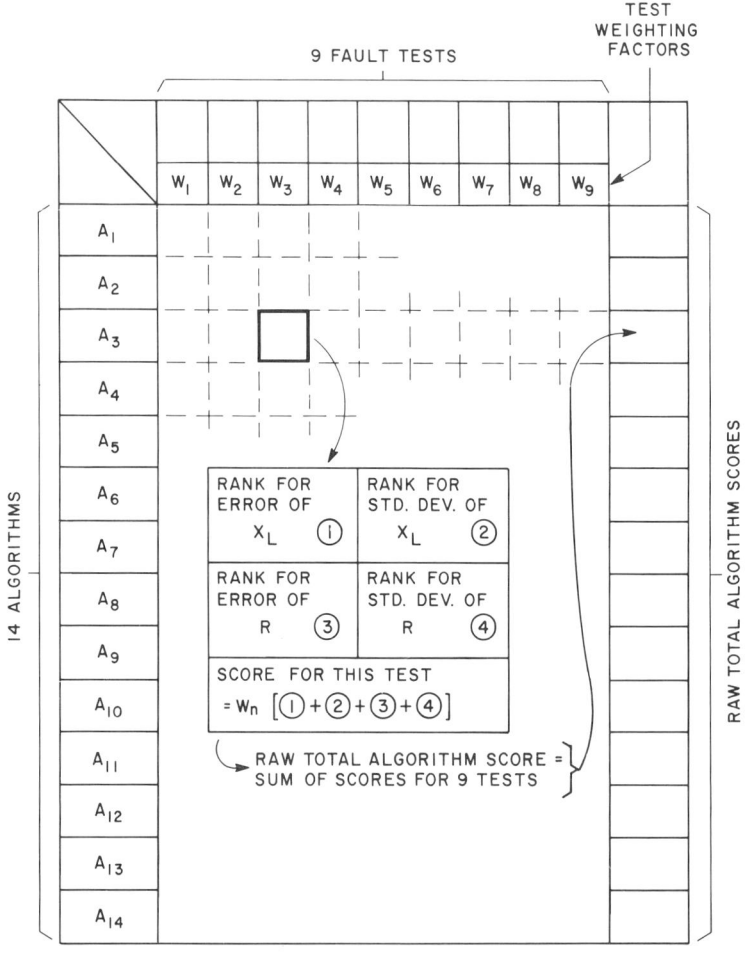

Fig. 9. Format of a Decision Matrix

X. RANKING FUNCTION

Percentage-error and standard-deviation results from the plots could not be directly used for ranking the algorithms. Obviously, if 0.5% error qualifies as good accuracy, it does not make sense to claim that 0.05% error is ten times as useful. Similarly, 50% error cannot be regarded as only half as bad as 100% error when neither is acceptable. Finally, the relationship between relative usefulness and percentage error values is not linear, even in the useful range. To account for these factors, the error was converted to a rank of between one and ten with the following function:

$$\text{Rank} = \text{Integer}\left[\frac{10\ \ln\ (LC/X)}{\ln\ (L/S)} + 1^{-}\right]$$

where X = percent error;
L = 25% for error of mean calculation, 50% for standard deviation calculation;
S = .75% for error of mean calculation, 3% for standard deviation calculation; and
C = 1 for reactance cases, and 2 for resistance cases.

L represents the boundary between a rank of 1 (barely useful) and 0 (useless). S is the value of error which lies at the top of rank 10 (so accurate that further improvement is not significant). Lower errors will produce ranks greater than 10, but such values are taken as 10. C is a scaling factor which expands the error tolerance for resistance calculation results. The use of C = 2 for the resistance errors reflects a value judgment made by the authors - that the reactance value is the more important component to consider when estimating fault location, and therefore that more error in resistance is acceptable.

The speed of response was evaluated using a similar formula based on a number of samples required for results to fall within a ±6% band for reactance data and ±12% band for resistance data using L = 32, S = 2, and C = 1. The percentages are defined on a base of expected values.

Figure 10 shows on a nomograph the ranks assigned to various percentage errors for the case of L = 25%, S = 0.75%, and C = 1, the values used with the reactance results. This agrees strongly with the investigators' intuitive evaluation of how increasing errors in the mean of X_L should be ranked.

Fig. 10. Ranks Versus Percent Error with
L = 25%, S = 0.75%, and C = 1

XI. ALGORITHM SCORES

For each algorithm and each test, a set of four ranks was produced; error and standard deviation for X_L and error and standard deviation for R. The four were added, and the sum multiplied by the weighting factor for that column as shown in Figure 9. For each algorithm, the results in the nine columns are added to form a raw score for that algorithm at the particular sampling rate and for the statistics beginning one-half or one cycle after the fault. Finally, the highest score possible was assigned the value 100 and all scores normalized accordingly. The results of this ranking process are given in the chart of Figure 11. The six graphs correspond to the six matrices used.

XII. CONCLUSIONS

On expected result is clear from visual inspection of Figure 11. Algorithms with long windows score low in the one-half cycle evaluations because of the delay in response. Those same algorithms, as a group, score best after one cycle because of their immunity to transients. Short-window algorithms show little change from the one-half cycle to the one cycle cases. Score differences of five to 10 points are not significant.

Although the procedures used to generate Figure 11 have reduced the volume of raw data to a manageable scale, new conditions must be considered to effectively utilize the results. Note that the algorithm having the highest absolute score requires a processor fast enough to handle 32 samples per cycle, and still does not yield an especially fast result. The authors imposed the further requirements of a high score after one-half cycle with lower, more practical sampling rates of 12 to 16 per cycle, all with minimum loss of relative performance in the one cycle cases. The McInnes-Morrison algorithm gives the best overall performance when these requirements are emphasized. It is important to note that such a conclusion is closely related to requirements set down by the authors at the outset of the study. A change in requirements may change the results - for example, reduced emphasis on speed combined with increased need for use of a low-cost processor might lead to the choice of a Fourier method with low sampling rate. The author of each of these algorithms has done an excellent job of satisfying the requirements he emphasized in his own work.

The original selection of the tests used is the dominant factor in the outcome of the reduction. In fact, the authors found that the scores are remarkably insensitive to substantial changes in the values of weights or ranking function parameters. A second evaluation was tried on one matrix with large changes in the weight and ranking parameters; a whole new set of results were expected. Surprisingly, the only change was that two adjacent algorithms in the middle of the ranking list were transposed.

There still remain factors which are totally outside the capability of the matrix technique. Some of these were brought to bear above in the selection of the McInnes-Morrison algorithm [12]. A more extreme example in the work described here is that the investigators were unable to quantify the results of the transformer inrush tests and have considered them separately.

The authors are gratified and reassured to observe that the Prodar 70 algorithm ("Rockefeller, et. al." in Figure 11), which did not score especially well in this ranking, has performed nearly flawlessly in six years of field trials. Newer, higher-scoring algorithms can only do better.

The user of any of these algorithms should realize that the performance of a specific method depends heavily on two additional factors - analog or digital pre-filtering of data samples, and the process by which the output values are converted into a relaying decision. Either or both may be used to overcome specific weaknesses of the algorithm. Since the possibilities for filters and for decision-making procedures are infinite, the user must apply engineering judgement here. The authors of this article are experimenting with selected pre- and post-algorithm processing methods for the most promising algorithms.

122 J. G. GILBERT et al.

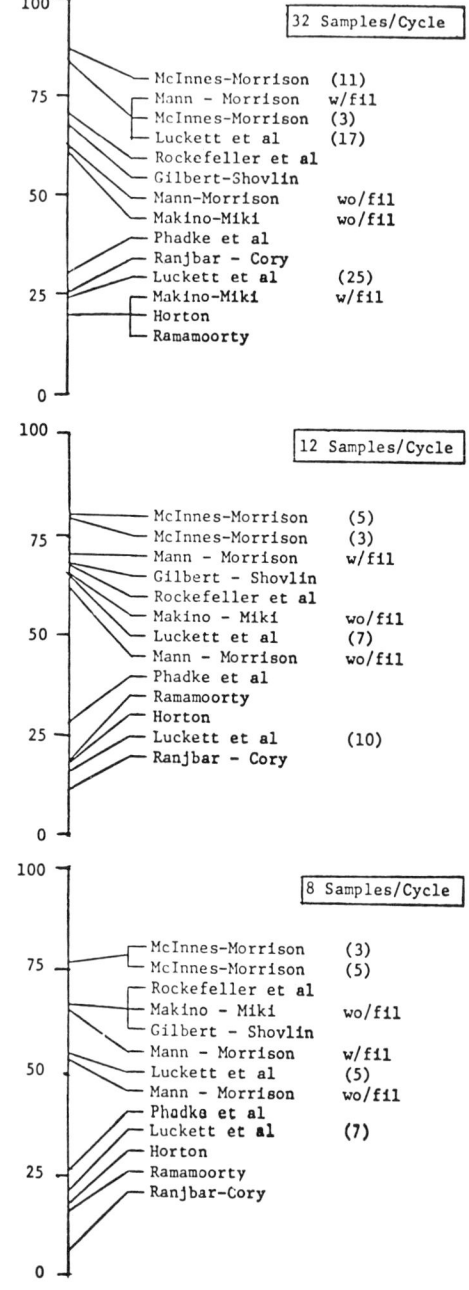

Fig. 11. Normalized Algorithm Scores (Numbers

DEVELOPMENT AND SELECTION OF ALGORITHMS

Data Produced By Algorithms

Beginning 1 Cycle After Fault Initiation

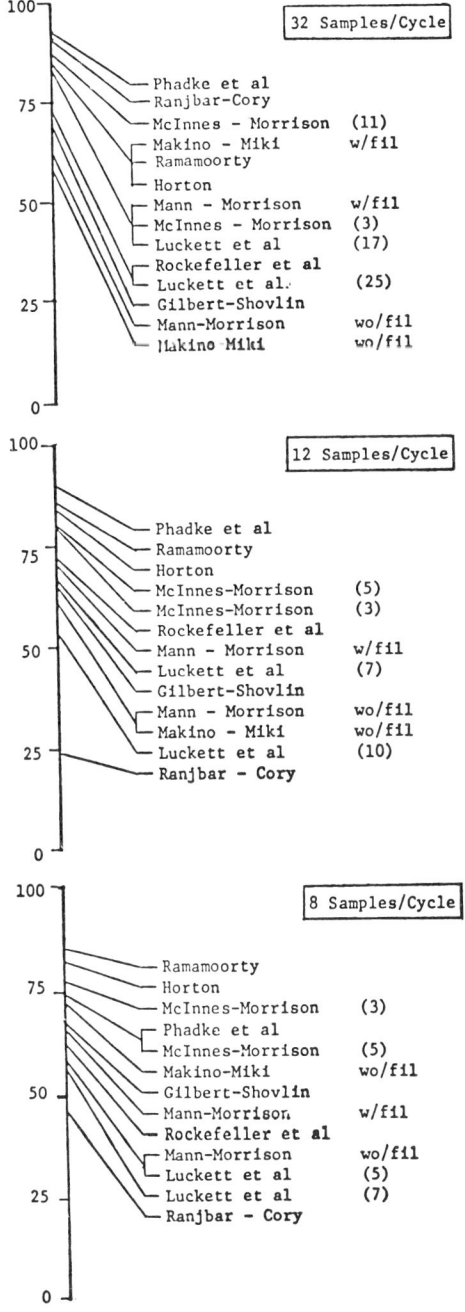

in parentheses are widths of data windows)

The combination of analytic data tests on computer relaying algorithms with matrices for reduction of the bulk of data produced has a number of idiosyncrasies and limitations, some of which this paper has identified. Nonetheless, the authors view it as an invaluable means of applying evaluation criteria in a uniform and objective manner to compare methods which have not previously been the subject of any universal comparison.

Finally, the authors feel that the algorithms tested here, or closely related variations, will serve as the basis for the bulk of commercially developed computer relaying when it is introduced. Other newer, more complex techniques, such as traveling wavefront analysis, will not supersede them in the near term, but will serve in specific applications where required.

XIII. ACKNOWLEDGEMENTS

Many individuals at Pennsylvania Power and Light Company and Westinghouse Electric Corporation deserve recognition for their advice and guidance. The authors thank Dr. C. W. Einolf of the Westinghouse Research Laboratories for producing the analytic data generation program.

XIV. REFERENCES

(1). Rockefeller, G.D., "Fault Protection with a Digital Computer", IEEE Transactions on Power Apparatus and Systems, vol. 88, April 1969, pp. 438-464.
(2). Mann, B.J. and Morrison, I.F., "Digital Calculations of Impedance for Transmission Line Protection", IEEE Transactions on Power Apparatus and Systems, vol. 90, no. 1, February 1971, pp. 270-279.
(3). Mann, B.J. and Morrison, I.F., "Relaying a Three-Phase Transmission Line With a Digital Computer", IEEE Transactions on Power Apparatus and Systems, vol. 90, no. 2, April 1971, pp. 742-750.
(4). Gilcrest, G.B., Rockefeller, G.D., and Udren, E.A., "High-Speed Distance Relaying Using a Digital Computer; I - System Description" and "II - Test Results", IEEE Transactions on Power Apparatus and Systems, vol. 91, no. 3, May/June 1972, pp. 1235-1258.
(5). Udren, E.A., "Relaying by Digital Computer - Experimental Field Trials and Future Prospects", EEI Systems and Equipment Meeting, Los Angeles, California, February 18, 1976.

(6). Phadke, A.G., Hlibka, T., and Ibrahim, M., "A Digital Computer System for EHV Substations: Analysis and Field Tests", IEEE Transactions on Power Apparatus and Systems, vol. 95, no. 1, January/February 1976, pp. 291-301.
(7). Ramamoorty, M., "A Note on Impedance Measurement Using Digital Computers", IEE-IERE Proceedings, vol. 9, no. 6, November/December 1971, pp. 243-247.
(8). Carr, J. and Jackson, R.V., "Frequency-Domain Analysis Applied to Digital Transmission Line Protection", IEEE Transactions on Power Apparatus and Systems, vol. 94, No. 4, July/August 1975, pp. 1157-1166.
(9). Laycock, G.K., McLaren, P.G., and Redfern, M.A., "Signal-Processing Techniques for Power System Protection Applications", IEE Conference Publication 125, Developments in Power System Protection, London, England, March 1975.
(10). Horton, John W., "The Use of Walsh Functions for High-Speed Digital Relaying", IEEE Paper No. A75 582-7, Summer Power Meeting, 1975.
(11). Hope, G.S. and Umamaheswaran, V.S., "Sampling for Computer Protection of Transmission Lines", IEEE Transactions on Power Apparatus and Systems, vol. 93, no. 5, September/October 1974, pp. 1522-1533.
(12). McInnes, A.D. and Morrison, I.F., "Real-Time Calculation of Resistance and Reactance for Transmission Line Protection by Digital Computer", E.E. Transactions, Institution of Engineers, Australia, EE7 no. 1, 1971, pp. 16-23.
(13). Ranjbar, A.M. and Cory, B.J., "An Improved Method for the Digital Protection of High Voltage Transmission Lines", IEEE Transactions on Power Apparatus and Systems, vol. 94, no. 2, March/April 1975, pp. 544-550.
(14). Gilbert, J.G. and Shovlin, R.J., "High-Speed Transmission-Line Fault Impedance Calculation Using a Dedicated Minicomputer", IEEE Transactions on Power Apparatus and Systems, vol. 94, no. 3, May/June 1975, pp. 872-883.
(15). Makino, J. and Miki, Y., "Study of Operating Principles and Digital Filters for Protective Relays with Digital Computer", IEEE Paper No. C75 197-9, Winter Power Meeting, 1975.
(16). Luckett, R.G., Munday, P.J., and Murray, B.E., "A Substation-Based Computer for Control and Protection", IEE Conference Publication 125, Developments in Power-System Protection, London, England, March 1975.

(17). Rader and Gold, <u>Digital Processing of Signals</u>, McGraw-Hill, New York, 1969.
(18). Ragazzini and Franklin, <u>Sampled Data Control Systems</u>, McGraw-Hill, New York, 1958.
(19). Kepner, C.H. and Tregoe, B.B., <u>The Rational Manager</u>, McGraw-Hill, New York, 1965.

Power Control System and Protection

FAULT CURRENT LIMITERS

Harry J. King

Hughes Research Laboratories

Richard E. Kennon

Electric Power Research Institute

I. INTRODUCTION

The prospective fault current levels in many power systems continue to rise as generation is added and the systems become more tightly integrated to improve system stability under various operating conditions. In many instances installation of fault current limiters (FCLs) at strategic locations may be the most cost-effective way of accommodating the higher fault levels without compromising system performance.

As with any new technological field, a great deal of work in various disciplines will be required to fully understand the function of an FCL in a power system and to optimize the performance and the cost/benefit ratio. There is a need to agree upon a common terminology as a basis for setting industry standards and for accurately disseminating technical results among interested parties. This article begins such an endeavor by reporting on the field of FCLs as it exists in the United States in mid-1977.

The basic need for an FCL was presented by Falcone et al.(1)(2) who used the American Electric Power Corporation's System as a model to quantitatively predict the rise in prospective fault currents during the 1970s and 1980s, and to estimate the cost of replacing breakers to handle these faults. These authors also defined the baseline criteria that an FCL must need to function satisfactorily. Hardware progress was reported by several authors at the IEEE Summer Power meeting in 1973.(3) More recently, a three-day "Symposium on New Concepts in Fault Current Limiters and Power Circuit Breakers"(4) sponsored by EPRI and the State University of New York was held in 1976 to provide a forum for workers in the field. Refer to the proceedings of the above symposium for a more complete description of the material summarized in this article.

II. BASIC FUNCTIONS OF A FAULT CURRENT LIMITER

Since a fault current limiter is a series device, it must present a low impedance to current flow under normal conditions. When a fault occurs, this impedance must rapidly increase to limit the current flowing into the fault. Conceptually, all types of FCLs may be viewed as a normally closed switch in parallel with a resistor.

The type of switch, its control circuit, and the type of resistor may vary widely from one design to the next. A dc switch may be used, which transfers current abruptly into the resistor at a fixed time after the fault occurs regardless of the phase angle of the ac voltage. An alternative is an ac switch or fuse that increases impedance more gradually than a dc switch and therefore shares current with the resistor until the first current zero when full current transfer into the resistor is made. Finally, the FCL may be a series-parallel resonant circuit whose component values change as the fault current rises so as to present increased impedance to the circuit. Some resonant circuit FCLs also include a resistor in their design.

The resistor can be a normal metallic element that has a constant current-to-voltage ratio or a nonlinear resistor made of silicon carbide or zinc oxide with asymptotic characteristics approaching that of a Zener diode which has an almost constant voltage drop regardless of the current flowing through it. In either case they absorb relatively large amounts of I^2R power when inserted in a transmission line for even a few cycles. This power increment appears in the resistor as heat and must be dissipated after the system has returned to normal and the FCL is in its low impedance state. Both the impedance switching mechanisms and the resistor types are described and compared in Section IV.

A typical FCL sequence is shown in Figure 1. Note the different type of current profile generated during the first half cycle by the ac and dc switching techniques. The peak current during the first half cycle is governed by the speed at which the resistor is switched into the circuit and the impedance of the resistor (as a function of current, if it is nonlinear).

Several useful and important terms may be defined with reference to Figure 1.(6)

I_{pA} = Available peak prospective momentary fault current without an FCL in the circuit (kA, peak). See eq. (2).

I_p = Major loop peak current with FCL, varies with ac or dc switching in Figure 1. (kA, peak).

I_A = Available arcing or fault current without FCL. (kA, rms).

I_{SS} = Steady state or arcing current with FCL. (kA, rms).

In principle, the current flowing through the FCL after the first half cycle may be reduced to an arbitrarily low level

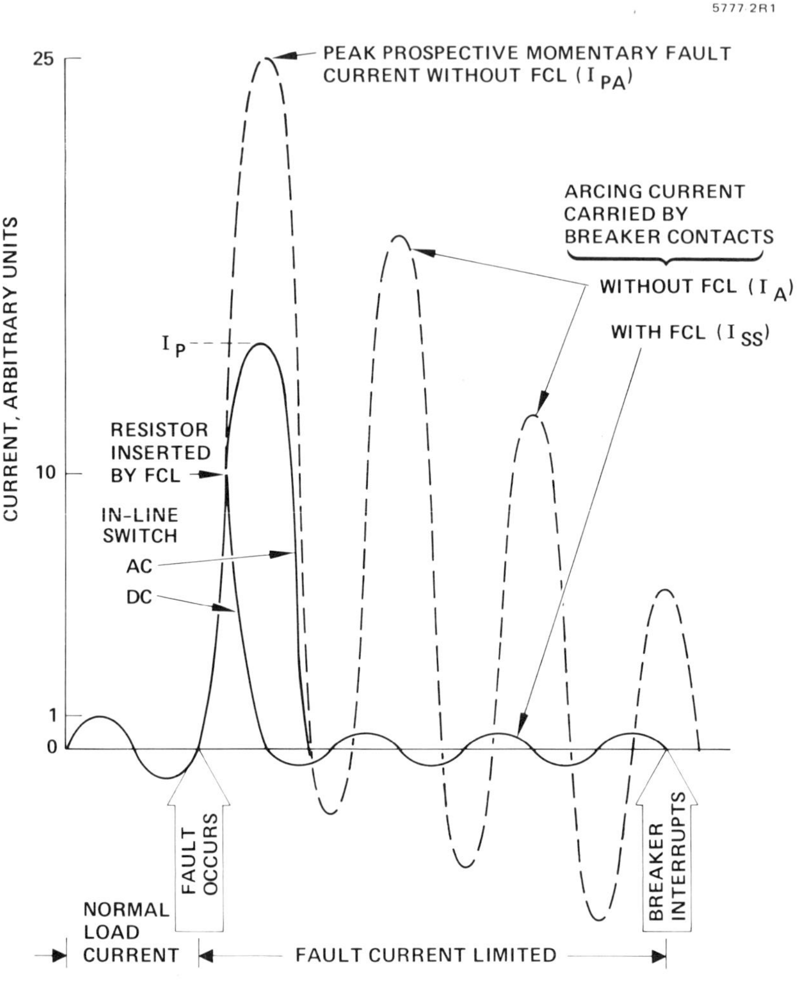

Fig. 1. The (prospective) fault current which would flow through a conventional breaker (broken curve line) as compared with the limited current which flows when the CLD inserts a resistor in approximately 1 ms (full curve).

by designing the FCL to have a sufficiently large impedance. In practice, there is an upper limit to the impedance which is given by

$$V = I_p \cdot Z(I) \tag{1}$$

where

V = maximum allowable system voltage set by insulation protective levels

I_p = maximum current flowing through FCL during fault

$Z(I)$ = terminal impedance of FCL as a function of current.

Note that the maximum system voltage is actually the vector sum of the voltage appearing across the FCL resistor plus the voltage across the line section from the FCL to the fault. Since the line is inductive, these two voltages add in quadrature, and the resultant is not much greater than that across the resistor alone. This is a primary reason for using a resistor rather than an inductor as the switchable current limiting element.

The nonlinearities associated with the intrinsic properties of the resistors, the thermal characteristics of the resistor material, current sharing between the resistor and the switching arc or the transient loop currents in resonant systems generally require an iterative computer computation to accurately predict the time variation of current and voltage during an FCL operation. Similar computations are required to calculate the amount of energy absorbed by the FCL.

A typical transient profile generated when a dc switch is used to insert a silicon carbide resistor with the nonlinear characteristic $V \propto I^{0.25}$ into a current carrying 6.5 kA is discussed in Section V. In this particular 138 kV system example, the energy absorbed by the resistor in each of the three phases during a 5-cycle insertion is 8 MJ. The voltage transient across the resistor at the time of insertion is 140 kV.

III. FAULT CURRENT LIMITER IN A POWER SYSTEM

Since power systems vary so widely in their design and operation, there can be no single FCL location that will be optimum for all systems. There are, however, several general application rules that may be illustrated with reference to Figure 2.

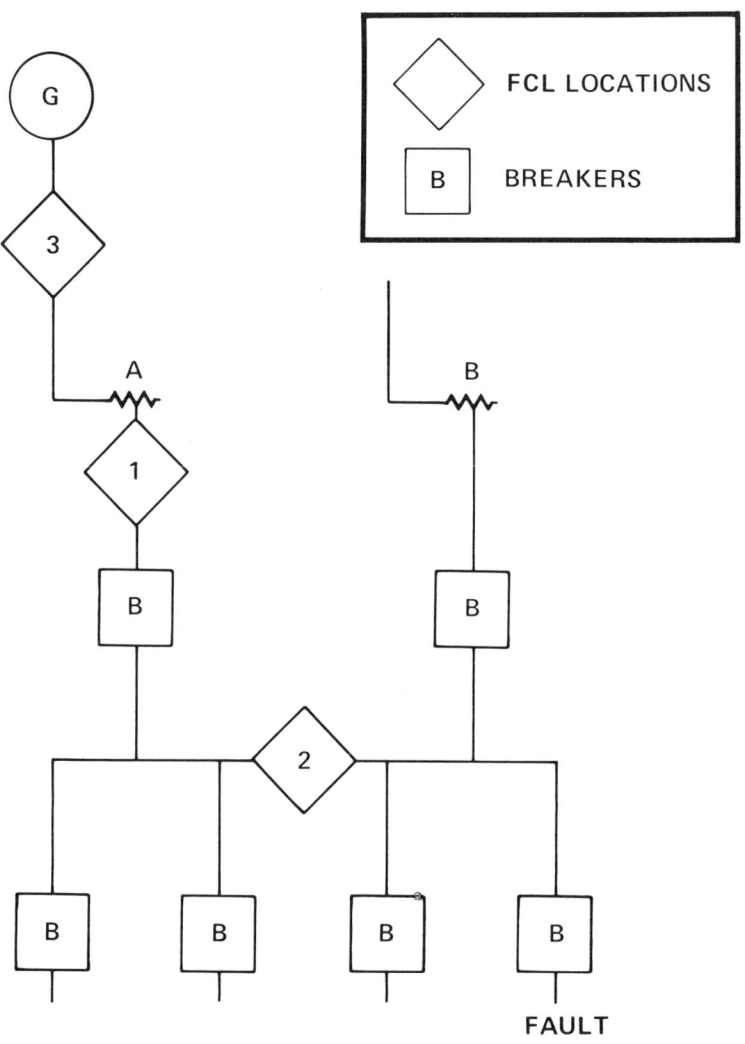

Fig. 2. System locations for FCL.

The first rule is that the FCL is not used to clear the fault but only to limit its magnitude so that a breaker can safely open without overload, as shown schematically in Figure 3. It is important that the fault current not be so severely limited that the breaker relaying system be defeated. Gallen(5) has concluded that this should be generally possible without major perturbation to existing relaying techniques.

Second, the FCL should be located so that it protects as many breakers as possible. The economic benefits of such a location are obvious. Locations 1 and 2 in Figure 2 both illustrate this approach. When used in position 1 (FCL 2 and 3 omitted), the fault contribution from transformer A is reduced or eliminated when the FCL is activated. The duty on the line breakers is decreased because the fault current contribution from the line containing transformer A is both reduced in magnitude and resistive in phase while the contribution through transformer B remains inductive. When used in the bus tie position 2 (FCL 1 and 3 omitted), the bus is effectively split and a faulted line can only be fed by transformer A or B, but not both. In real systems, the buses are more complicated than shown here, and in general, two FCLs are required to effectively split the bus. An FCL at the generator output (location 3) is typical of one of the more unique system locations that has been suggested (2). While this situation has not been fully analyzed, it may prove possible to both protect the generator and provide some phase stability while switching is accomplished.

Third, at least when located in a source (location 1 or 3 in Figure 2), the FCL need not be sized to switch the entire current at the fault location, but only in its own source. Elimination of the fault contribution from one source is often all that is required to prolong the service of a number of line breakers on a bus. This generally means that although the most obvious FCL location is in the bus itself, it may be more cost effective to place the FCL in one or more source lines to reduce the total bus fault while leaving the bus electrically intact.

Because of the unique way in which FCLs are used in a power system (i.e., they are usually not put in the faulted line but in series with a source which feeds it), there is a possibility of confusion in discussions between the system designer whose breakers are to be protected and the FCL supplier whose device is to be specified. Note that the currents and current ratios through the FCL are not the same as those through the breakers and that the two can only be related by a systems analysis.

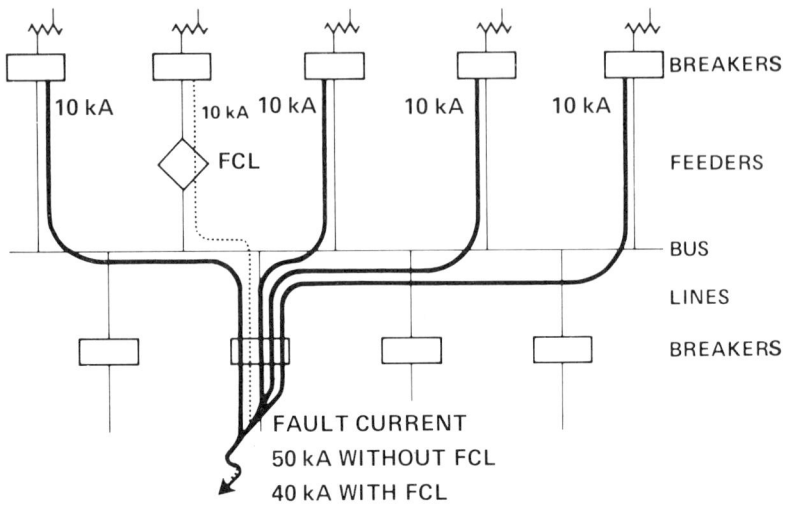

Fig. 3. An illustration of how a feeder line with 10 kA prospective fault current capability may be added to a bus without materially increasing the fault duty on the line breakers.

With the above general rules in mind, the procedure for specifying an FCL may be briefly defined as follows:

1. On a system one-line diagram, define the sources of current which flow to the point of the fault.

2. Identify the location of the breakers that are overloaded and the magnitude of the overload under worst case fault conditions.

3. Check to see if the elimination or reduction of the fault current contribution from any one source line or bus will reduce the fault current through the overloaded breakers to an acceptable level. If so, this is a possible FCL location.

4. Verify that location of an FCL at this position will not present relaying problems, stability problems, or physical problems within the station.

5. Once a tentative location is chosen, the preliminary specifications necessary for a manufacturer to provide a rough cost estimate are:

    a. Line voltage

FAULT CURRENT LIMITERS

    b. Normal and overload continuous current ratings

    c. Prospective fault current throughput

    d. Peak allowable current when FCL is required to operate

    e. Maximum steady state limited current with FCL in system

    f. Peak voltage when resistor inserted

    g. Recycle time (resistor cooldown time)

    h. Physical details (indoor, outdoor; live tank, dead tank, etc.)

    i. Number of operations before maintenance

    j. Expected breaker clearing time

    k. Redundancy required.

6. The economic tradeoff must compare the installed cost of the FCL against the installed cost of the new breakers, bus work, and switches that may be required if an FCL is not used. This analysis should also include the real but less tangible costs and risks which are associated with operating a split bus or working in and around an existing station which is an integral part of an operating system. There is also a potential cost saving due to the reduction in transformer through faults when FCLs are situated as in number 1, Figure 2.

7. Finally, there are a number of FCL specifications which are vital to its operation and successful integration into the power system, but which have only a second-order effect on its cost. Some of these are

    a. Threshold trip settings

    b. Sensitivity to fault current direction through FCL

    c. Reclosure sequence to be compatible with breakers

d. Interfaces with system controls and relaying

e. Protection of system against FCL malfunction.

Installation of an FCL in a system as described above should produce several system benefits. Most important, it should be cost effective because the useful life of existing breakers, switches, and buswork will be extended. Although transformers are designed to withstand heavy throughfaults, a reduction in these current peaks will reduce internal transformer stresses and should ultimately extend their life.

An FCL may be constructed in an existing station and then integrated into the station in a relatively short period of time. This is both convenient for scheduling and much safer from an operating standpoint, since the station will not be out of service or on limited service for extended periods.

IV. FAULT CURRENT LIMITER HARDWARE

A. Sensing

Sensing faults to determine whether or not a fault current limiter is to be operated becomes a special problem if one assumes that the first current crest must be limited. For instance, if the first current crest must be limited to half its maximum peak prospective value by an ac switch (Figure 1), there can be faults initiated with phase angle such that the switching sequence must be complete in less than 3 msec. When a dc switch is used, it is necessary to limit the peak current that is commutated into the resistor both to limit the voltage transient (eq. 1) and to stay below the switch current ratings. This typically requires a sensing time of 400 μsec and a switching time of 1.5 msec (Section V). Since these times are less than one quarter cycle at 60 Hz, the difficulty of discriminating between faults and other system disturbances becomes apparent. Figure 4(a) illustrates various wave shapes that a sensor might encounter. Furthermore, it is desirable to discriminate between faults which must be limited and faults which are already constrained by the source impedance to a low enough magnitude not to require limiting. Several parameters connected with these wave shapes which might be utilized for sensing a fault are

1. di/dt — rate of rise of current

FAULT CURRENT LIMITERS 137

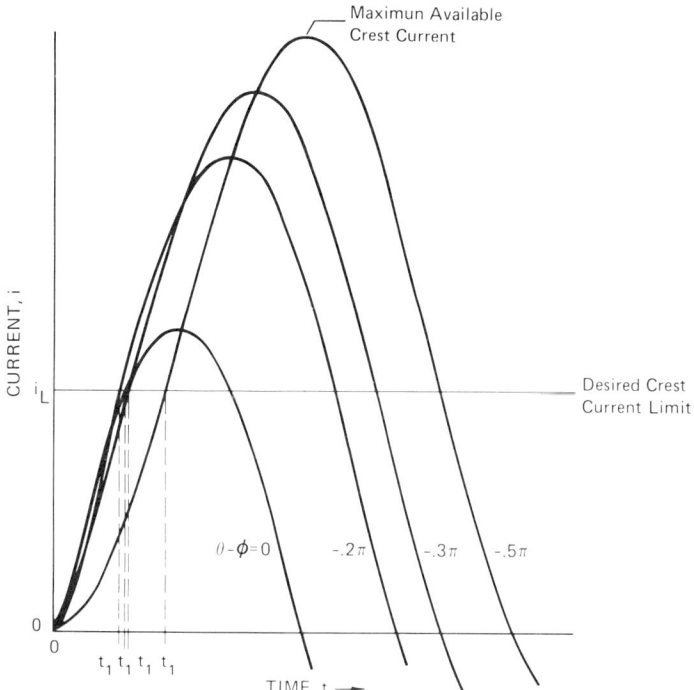

Current as a function of time — maximum available fault

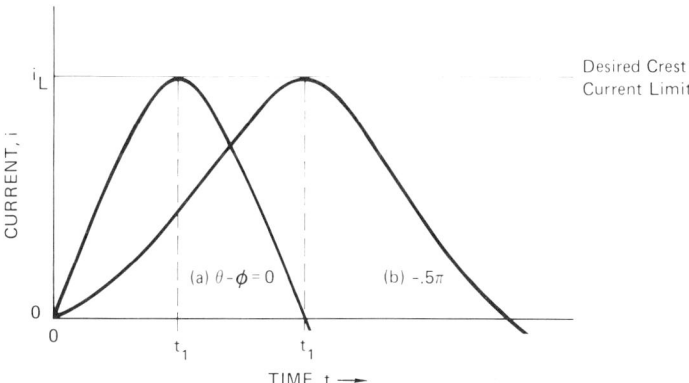

Current as a function of time — minimum fault requiring action

Fig. 4. Top figure — increase in peak momentary crest current as the fault becomes more fully offset (i.e., $\theta - \phi \to -\pi/2$). Bottom figure — difference in dI/dt possible between two faults of the same magnitude but differing $\theta - \phi$.

2. $|I|$ — current magnitude

3. I^2T — heating or energy function

4. ΔT — time duration

5. Comparison with reference fault.

Analysis of all possible wave forms for each particular application will yield these parameters for all of the fault initiation angles. The problem then becomes one of determining the range of these parameters for the range of faults to be detected. Second, it must be determined if these parameters are within the indicated range for a sufficient time to discriminate between faults and other system disturbances. Third, it is important to determine if these parameters can be measured and verified in sufficient time to allow for the operating time of the fault current limiter itself. For better discrimination, one might want to combine parameters. For instance, one might choose to issue a signal if both the di/dt and i magnitude are above certain threshold limits for a specified ΔT time. Studies along these lines for a particular application have been conducted by Gould, Inc. (6), (17), and it was determined that for the application in mind, a simple di/dt measurement with the requirement that the di/dt be above a certain threshold for 0.4 msec was sufficient to discriminate between faults and capacitor bank switching, switching surges, or surge arrester operations. Studies were not made to determine if this sensing criterion would distinguish between faults and transformer in-rush currents. However, for the application in question, it was determined that transformer in-rush currents were not a problem. Such a simple sensing scheme has the advantage of low cost and the other attributes that go with a simple device. Studies must be made for each application to determine if this approach can be utilized.

For analysis, consider the general fault current equation:

$$i = I_m \{\sin(\omega t + \theta - \phi) - \sin(\theta - \phi) e^{-R\omega t/X}\} \quad (2)$$

where

$$I_m = V_m/(X^2 + R^2)^{1/2}$$

θ = angle of voltage waveform at which fault is initiated

$$\phi = \tan^{-1} \frac{X}{R}$$

X, R = reactive and resistive components of impedance from source to fault location

V_m = crest of source voltage.

The first term inside the brackets gives the steady state solution and the second term the transient or decaying portion. The system parameters (V_m, X, R) define the maximum fault current peak possible under worst case conditions ($\theta - \phi = -90°$) and the rate that the fault current envelope will decay from the momentary offset current peak which occurs during the first half cycle. The variable ($\theta - \phi$) which is present in both terms defines the degree of asymmetry or offset of the fault current. The set of curves plotted in Figure 4 are representative.

The rate of current change for the fault current of eq. (2) is

$$\frac{di}{dt} = I_m \omega \{\cos(\omega t + \theta - \phi) + \frac{R}{X} e^{-R\omega t/X} \sin(\theta - \phi)\} \quad (3)$$

which is also complex and contains both steady state and transient terms.

Several strategies can be used for ensuring an operation for any prospective crest current level above the desired current limit yet minimizing the probability of operation for fault currents below the limit. The problem stems from the fact that the maximum di/dt for curve (a) Figure 4b is approximately double the maximum di/dt for curve (b). If one sets a di/dt sensor to operate at the level of the maximum for curve (b), then symmetrical faults (i.e., $\theta - \phi = 0$) with crest magnitude less than the desired limit will cause FCL operation.

It becomes the task of the designer to combine the system parameters, V_m, X, R with the desired current limit and with the capabilities of the FCL to determine the sensing criteria which will operate the FCL for every fault above the desired limit and a minimum number of faults below that limit. It is apparent that such a study will require iterative techniques with many scenarios tested on a digital computer simulation. Studies have led the authors to believe that no combination of the parameters, di/dt, $|I|$, and Δt are sufficient to guarantee no operations for prospective fault currents below the desired crest current limit with this type of sensing scheme.

A considerably more sophisticated and universally applicable approach(8) has been taken by the Hughes Research Laboratories who have developed a fault sensing scheme described in Section V which compares each line current with a reference signal to precisely estimate the severity of a fault during the first 400 μsec after its occurrence.

B. In-Line/Bypass Switches

It is essential that any fault current limiter be capable of carrying load current on a continuous basis without deterioration. Furthermore, it is certainly desirable that this load current carrying capability be accomplished with minimum losses. Any load current losses would have to be calculated in an evaluation of the device along with first cost, operating cost, and maintenance costs. Although there are many so-called high-speed switches available, in general they do not meet the speed requirements of a fault current limiter. Several new concepts have been proposed and developed in recent years. For most applications envisioned at present, the switch must be capable of operation in times of 0.5 to 1 msec. Such a speed requirement precludes the use of coil springs, linkages, latches, etc., as we know them from power circuit breaker technology. Furthermore, since the bypass switch will normally be required to commutate current from its own contacts to some parallel device, it is necessary that the bypass switch develop a reasonably high arc voltage within its total operating time limit. Once the current is commutated into the parallel device, the bypass switch must regain its dielectric strength at a rate exceeding the rate of rise of voltage across the fault current limiter. This requirement may be in the tens to hundreds of volts per microsecond.

Two categories of devices to be considered are single-shot devices and mechanical switches with multi-operation capability. An example of a single-shot bypass switch is that used by the Calor Emag Company in its I_S-Limiter (13) (14). In this design, the main current carrying member is a hollow tube which is scored near the middle of its length. Inside the tube at the location of the scoring is a powder charge. This conductor is then encased in a heavy walled insulating tube. When an operation is called for, the powder charge is detonated and the conductor parts in the middle. The advantages of speed are obvious and the device generates sufficient arc voltage to commutate the current into a parallel fuse element. It is necessary, of course, to replace this device after each operation. Other fast-acting single-shot chemically activated switches are under development. It is expected they will all require a certain amount of refurbishing after each operation.

A novel in-line mechanical switch capable of repeat operations has been developed by the Hughes Research Laboratories.[7] This device dumps the energy stored in a capacitor bank through an electrodynamic drive motor to open a set of contacts which are mounted on a torsion bar. Opening time is 1 msec. The contacts lock open until they are closed by a current pulse from another capacitor bank.

A tuned circuit type of fault current limiter requires a switching capability with entirely different specifications. In these types of limiters, the continuous current is carried by an inductive and capacitive element in series and a bypass switch protecting the capacitors is normally in the open position. Furthermore, the bypass switch is generally only closed for a short period and, therefore, requires only a momentary damping. A spark gap is the most commonly used "switch" to protect a series capacitor in a transmission line. Since the capacitor portion of a tuned fault current limiter is very similar to a series capacitor, it seems quite logical that the same type of bypass protection can be used. Gaps have been developed by the Westinghouse Electric Corporation which will carry 30,000 A rms for 30 cycles without significant deterioration of performance. Of course, if the gap is used in this application, a triggering scheme must be utilized. This seems to be a straightforward design problem, but to the authors' knowledge, it has not yet been done for the fault current limiter application. Tuned limiters developed in Europe (9) (10) use a saturable reactor for the capacitor bypass. This scheme has proved effective in a number of applications.

C. Interrupters/Current Commutators

With fault current limiters as we know them today, the bypass/in-line switches are capable of generating only enough voltage to transfer the current into a second low impedance device which has the capability of changing impedance rapidly, but does not have the capability of carrying continuous current.

The interrupter is required to carry current for a short period of time until the in-line device can clear and start building dielectric strength. It is then required to increase its impedance or, looking at it in another way, generate back voltage to the point where it either interrupts the circuit or successfully commutates the current into a parallel impedance. In either case, it is then required to withstand the voltage drop across the fault current limiter until such time as it is again bypassed.

For instance, referring again to the Calor Emag I_s-Limiter, this second device is a silver-sand current limiting fuse. Once the current is transferred into the fuse, the element can heat to melting or bursting in very short time periods of the order of 0.5 to 1.0 msec, and in the process generate considerable arc voltage. This arc voltage is sufficient to limit the current to the prescribed level, and either interrupt current or transfer it into some other parallel high impedance element.

This same concept is being used by Gould, Inc. in their design of a 69 kV transmission fault current limiter for application on the Southern California Edison Company system. This is being done under EPRI Project No. RP281. Another type of interrupter which will be used to commutate current into a parallel resistance element is the crossfield interrupter being developed by Hughes Research Laboratories. This device (7) is basically a cold-cathode, glow discharge device. Current conduction depends on the presence of crossed electric and magnetic fields. When the magnetic field is removed, conduction ceases and voltage holdoff capability rises at about 5 kV/μsec up to 100 kV.

Another device utilizing somewhat similar technologies is a vacuum plasma switch being developed by the State University of New York at Buffalo (SUNYAB). The device consists of a centrally located cathode with an axial anode of larger diameter as shown in Figure 5. Arcing is initiated by a trigger electrode near the cathode and the device can be caused to conduct current between cathode and anode with voltages as low as 50 V across it. When an axial magnetic field is applied by the coil near the anode, the impedance can be made to increase by many orders of magnitude. Thus, the current flowing in this device can be forced to commutate into a parallel impedance. By properly controlling the magnetic field, this device can be further refined to serve as a limiter instead of an interrupter. Since the energy would be absorbed into the anode, it must be dimensioned accordingly. It appears that such a configuration might be feasible for distribution voltages of 15 kV and below and fault currents below 10,000 A. Work is being done on this device at SUNYAB, McGraw-Edison

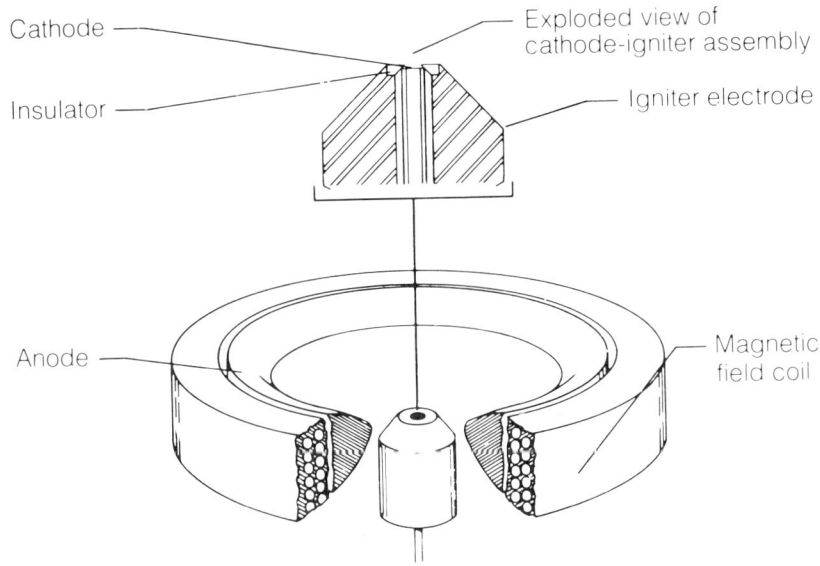

Fig. 5. Vacuum arc device capable of limiting and interrupting the flow of current.

Laboratories in Franksville, Wisconsin, and Gould, Inc. in Greensburg, Pennsylvania.

It would be inherently advantageous if the in-line/bypass switch could be combined in the same device with the interrupter. One attempt at doing this is being carried out under EPRI Project No. RP564 by the Westinghouse Electric Corporation Research Laboratories in Churchill Boro, Pennsylvania. It has been found that a vacuum interrupter such as those used in circuit breakers can be made to develop considerable arc voltage if a transverse magnetic field is applied to the arc plasma after the electrodes have been separated. It has been found necessary to use a resistor/capacitor network in parallel with this device to achieve commutation. Schematically, this device and its operating sequence is shown in Figure 6. It has been demonstrated that currents up to 10,000 A can be successfully commutated into the parallel network. Work is continuing to increase this limit and to make a single device capable of operating in a 69 kV fault current limiter. Again, the device must be opened to a specified dimension and the magnetic field applied in times of the order of 1 to 2 msec.

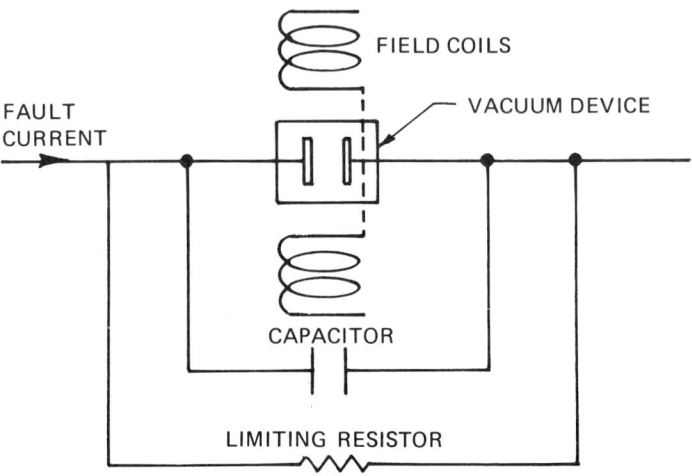

Fig. 6. Schematic of FCL of dc switching type using force commutated vacuum switch.

Another possibility for combining the continuous current carrying member with the interrupter is a superconductor. In the superconducting state, the materials exhibit zero resistance and small amounts of material can carry large currents continuously. Then the material can be caused to go "normal" by either an increase in temperature or an increase in magnetic field beyond a critical value or a combination of the two. Work has been done to evaluate such a system on EPRI Project RP328 by the Argonne National Laboratories (15).

D. Energy Absorbers

Switched impedance fault current limiters require an impedance in parallel with the switching means to limit current until a circuit breaker clears the fault. Most designs under active development propose to use a resistor for this application. Advantages are that a resistor will absorb energy stored in the source inductance, make circuit interruption easier and fault current contributions from inductance limited sources will add in quadrature to the resistance limited current.

Resistors for this application can be either linear or nonlinear. Linear resistors (current is proportional to voltage) can be made from any number of well-known materials, both metallic and non-metallic. Factors to be considered in resistor design are total energy absorption, rate of energy absorption, temperature limits, magnetic forces, expansion and contraction, voltage withstand, and cooldown time. While resistor technology is well known, there are not many applications requiring the very large surge energy absorption capability of fault current limiters. For instance, the 69 kV limiter to be installed at Southern California Edison must absorb 89 MJ per phase with the specified duty cycle. All of this energy must be absorbed in the resistance material, as there is not enough time for significant heat transfer. The resistor for this application will be made of metal tubing filled with magnesium oxide, much like range top cooking units. This tubing will be wound in pancake coils and stacked one above the other and connected to minimize inductance. The overall size of the resistor assembly will be 3 ft in diameter and 6 ft in height.

The Bonneville Power Administration has developed a braking resistor which has similar requirements and has been reported (11). Review of this effort is instructive as it points out a number of design problems presented by this type of application.

Another linear resistor candidate is ceramic bonded carbon. The resistors come in many forms, such as solid rods, hollow tubes, or short, large diameter solid cylinders. For most applications, it will be necessary to connect many units in series/parallel. The short, solid cylinders are the most economical and probably the most easily assembled into a large series/parallel grouping. Such an assembly will be compact and inherently low inductance. The principal disadvantage will be cooling time. Of course, there are relatively simple ways to reduce cooling time if that is a problem.

Liquid electrolyte resistors do not seem practical to the authors as the rate of energy input tends to cause localized vaporization and explosion. Another proposed design concept is a saturable reactor with thick steel laminations to make use of eddy currents (12). It is anticipated that such a resistor, while inductive, could have a power factor of 0.89. The prospect of absorbing energy in ordinary magnetic steel would appear to have cost advantages. This type of resistor would be highly nonlinear.

The more commonly available nonlinear resistor is made of silicon carbide particles bonded together in a ceramic matrix. The current through a silicon carbide nonlinear resistor is proportional to a power of the voltage such as three to five. This type of resistor is inherently voltage limiting; thus a surge voltage produced by the FCL can be easily limited to a reasonable level. These resistors are available in the form of short, large diameter cylinders. Thus they possess all the advantages of the bonded carbon resistors. An application of silicon carbide nonlinear resistors will be described later. Another nonlinear material under intensive development is doped zinc oxide. This material can be made with exponents an order of magnitude larger than silicon carbide with the same or better energy absorbing capability. Also silicon carbide has a negative temperature coefficient of resistance whereas zinc oxide is stable over a wide temperature range.

Most, if not all, fault current limiters must absorb some energy, even if only to damp oscillations. This is especially true of tuned limiters. Under some conditions, undamped tuned limiters can cause such a large offset of current that some current zero interruption opportunities will be missed.

The switched impedance limiters using silver-sand fuse elements absorb considerable energy in the fuse element itself. Under some conditions, the fuse element in the Southern California Edison prototype will absorb approximately 1 MJ. Applications described (13) for the Calor-Emag I_S-Limiter indicate that all energy absorption is done in the fuse.

It is difficult to classify one additional working FCL type in the same terms used for the others in this paper. This is the saturable reactor type installed on the Tokyo Electric system (16). The principle of operation is based on the fact that the inductance of a saturable reactor is low when the core is saturated and high when it is not. It undoubtedly has the advantage of a static device that does not require a separate sensor. Losses must be evaluated as part of the cost just as they are with a series tuned limiter.

Table 1 has been prepared to summarize all of the types of FCLs either in use or being developed that are known to the authors.

V. THE HUGHES/AEP SWITCHED RESISTOR FAULT CURRENT LIMITER

A fault current limiter is being installed by the Hughes Aircraft Company in the Muskingum River Switchyard of the American Electric Power/Ohio Power Company. The 345 kV/138 kV switchyard, which is not at present in danger of overload, was chosen as a convenient site for a first installation because its configuration is typical of other larger yards in the AEP system, it is conveniently located, and the maximum fault levels are consistent with the interruption capabilities of

Table 1

Developer	Sensor	In-Line Device	Interruptor or Commutator	Energy Absorber	Voltage Class, kV	Reference
Hughes	Simulation	Mech. Switch	X Field Tube	SiC	69-345	Section V
Gould, Inc.	di/dt	Mech. Switch	Silver-Sand Fuse	MgO Filled Inca-oy Tube	34-242	(5), (17)
SUNYAB	NI	NI	Fixed Electrode Vacuum Plasma	NI	15-69	(18)
Westinghouse Vac	NI	Movable Electrode Plasma Switch	Vacuum	NI	15-69	(19)
Westinghouse Tuned	Voltage	Tuned Circuit	Spark Gap	Linear Resistor	242-500	(20)
English Electric	Saturable Reactor	Tuned Circuit	Saturable Reactor	Linear Resistor	15-500	(8)
Calor Emag (B-V)	di/dt I	Scored Al Tube	Silver-Sand Fuse	Fuse	4-34	(12)
Tokyo Electric	----------	---------- Saturable Reactor ----------			69-242	(15)

NI = Not included in project

the available current limiter components. Design of the first device was begun in 1974, on-line fault sensor tests were completed in March 1977, and final installation of the complete system is scheduled for early 1978.

A. Power System

A simplified one-line diagram of the station is shown in Figure 7. The FCL is placed in series with the 150 MVA "A" feeder transformer. When activated, the FCL removes the fault current contribution that flows through transformer A to the bus and thus reduces the interrupting duty on the line breakers (such as "Natrium") by 25 to 50%, depending on the exact bus configuration at the time of the fault. Figure 7 also shows the location of the monitoring station (MS), which was installed to test the prototype fault sensor and to record the line transients that occur during the operation of the FCL. Note also the ability to isolate the two buses and the initial 13 mile section of the Natrium line, both convenient features during the installation and checkout of the FCL when a series of staged faults are required. Table 2 lists the prospective maximum fault current for faults occurring at the bus and on the Natrium line.

Figure 8 illustrates the FCL operation for a bus fault which is the worst case fault that can occur on the system. Parametric studies indicate that the current to be interrupted rises most rapidly for a phase to ground fault which occurs near the peak line voltage. It may be seen from Figure 8 that when the maximum interrupting capability of the FCL switch is 7.0 kA, there is only 1.3 msec available to recognize the fault and complete the switching operation. This is the design basis for the installation.

Resistor insertion in 1.3 msec will reduce the peak momentary fault current through both transformer A and the line breaker in series with the fault. It will not disturb the relaying around the faulted section because contributions from transformer B and the generators will still flow into the fault.

B. Fault Sensing

The mechanical switching operation requires 900 μsec. Deducting this opening time from the 1.3 msec maximum resistor insertion time leaves 400 μsec for the control system to recognize that a severe fault has occurred and to issue the command

FAULT CURRENT LIMITERS

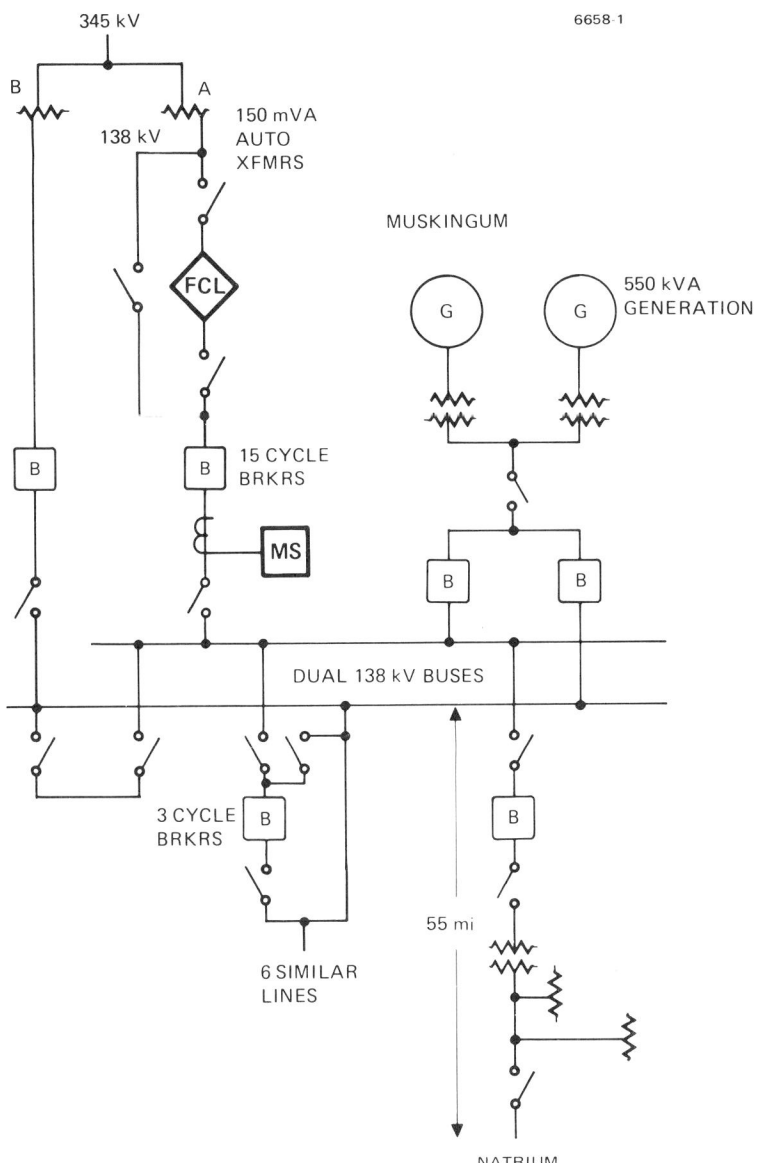

Fig. 7. Fault current limiter (FCL) and monitor station (MS) in Muskingam River 344 kV/138 kV Ohio Power Switchyard. Also shown is 55 mile line to Natrium used for tests.

Table 2. Muskingum Switchyard — Fault Currents

	Current through FCL kA, rms sym			Current at Fault kA, rms sym		
	ℓ-g	ℓ-l	3ϕ	ℓ-g	ℓ-l	3ϕ
BUS FAULTS						
1. Both XFMRS A&B in service	6.7	4.3	5.0	31.5	23.2	26.4
2. XFMR B out of service	7.5	4.7	5.4	26.1	19.5	22.4
NATRIUM LINE TEST						
1. At 10 miles	5.0	—	—	5.0	—	—
2. At 18 miles	2.8	—	—	2.8	—	—
3. At 48 miles	1.2	—	—	1.2	—	—

to begin the switching operation. Nuisance tripping due to low-level remote faults, lightning strikes, or switching transients should be reduced to a minimum. Specifically, it was decided that the FCL should operate only on true faults that produce a peak momentary current at the FCL location which is greater than 5 kA.

The fault sensor developed (8) for this installation accommodates the electrical phase angle (θ, eq. 2) at which the fault occurs in the following way. Consider first a single-phase system in which an electrically isolated, scaled (i.e., 1 kA = 1 V) line current signal is available. A simulated fault current signal with a peak value of 5 V (equal to a 5 kA fault) is continuously generated at a phase angle determined by the calculated (X/R) value of the fault current ($\phi \approx 80°$). A peak threshold current signal of 1.5 V (equal to 1.5 kA) is preset into the control.

Operation is as follows. When the line current is less than the threshold current, the system is quiescent. When the 1.5 kA threshold is exceeded, the function

$$F = \int_0^{400 \; \mu sec} (I_{LINE} - I_{5K \; FAULT}) \, dt \tag{4}$$

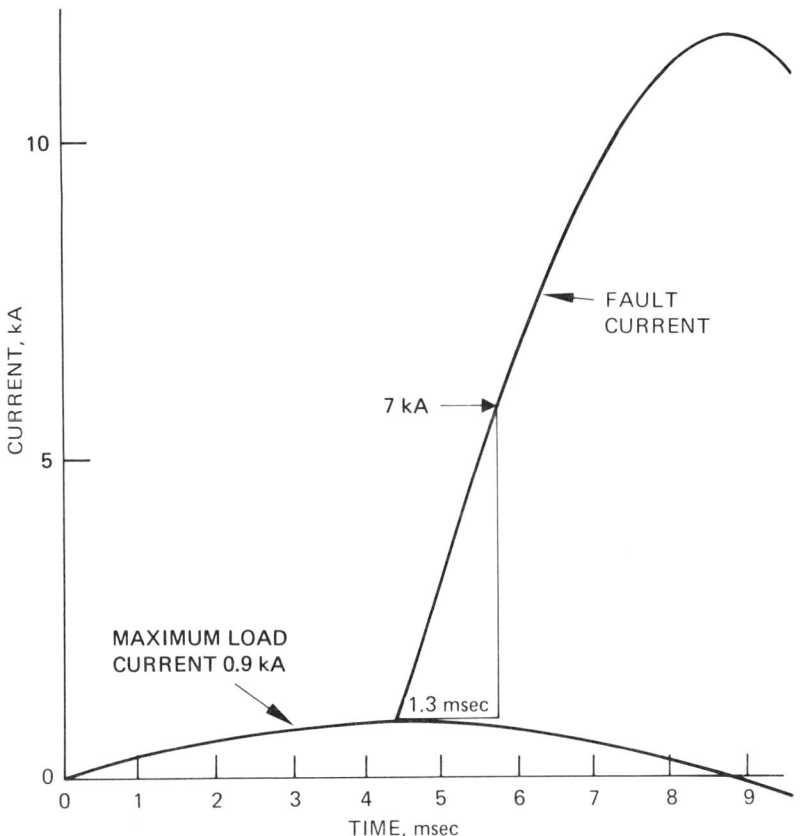

Fig. 8. Relationship between FCL operating time and current at the time of resistor insertion (for limiting case ($\theta - \phi = 90°$)).

is generated. If, after 400 μsec F is negative, either the transient was very short or of amplitude less than 5 kA and the FCL should not be actuated. If, however, F is positive, then a large sustained current rise is indicated and the signal to insert the resistor is given. This system is readily extended to three phases by generating three, rather than one, reference fault signals and incorporating the necessary logic to interpret the results. It recognizes line to ground faults by the above method, line to line, 3ϕ, and evolving faults by the di/dt method described earlier and gives commands for single pole switching for ℓ-g faults and 3ϕ switching for all others. Table 3 illustrates the consistency of the control system performance when evaluated on an analog simulator where faults were initiated at various angles.

Table 3. Tests of Fault Sensor on Simulator. Note lower prospective fault current setting required for line to line (di/dt) sensor.

Fault Phase Angle, deg	Fault Type, kA (peak)			
	Line to GND		Line to Line	
	$I_{Prospect}$	$I_{Resistor}$	$I_{Prospect}$	$I_{Resistor}$
0	5.2	3.7	3.3	4.2
40	5.2	3.9	3.3	3.2
90	5.4	3.7	3.3	3.2
140	5.3	-3.8	3.3	4.2
180	5.3	-3.7	3.3	4.0

$I_{Prospective}$ = Peak symmetrical fault current that would be reached with no FCL action.

$I_{Resistor}$ = Current through resistor at moment of resistor insertion; negative values indicate resistor insertion in second half cycle after minor loop.

C. Power Switching Circuit

The power circuit of the HAC FCL is shown in Figure 9. Normally the mechanical in-line switch contacts are closed and present essentially zero impedance to the power flow. When the command is given to insert the resistor the in-line switch opens, generating an arc voltage of 1.5 kV in 700 μsec.* The interrupters are triggered and rapidly (< 50 μsec) commutate the current out of the in-line switch. The current is held in the interrupters for 200 μsec to give the in-line switch a chance to fully open and to deionize. The interrupters cease to conduct after 200 μsec and the current is then forced into the resistors creating a maximum voltage kick of 70 kV

---

*Switch Specifications

    Arc Voltage Rise   = 1 kV/msec

    Dielectric Recovery   = 10 kV/μsec

    Rate of Current Change at Interruption   = 50 A/μsec

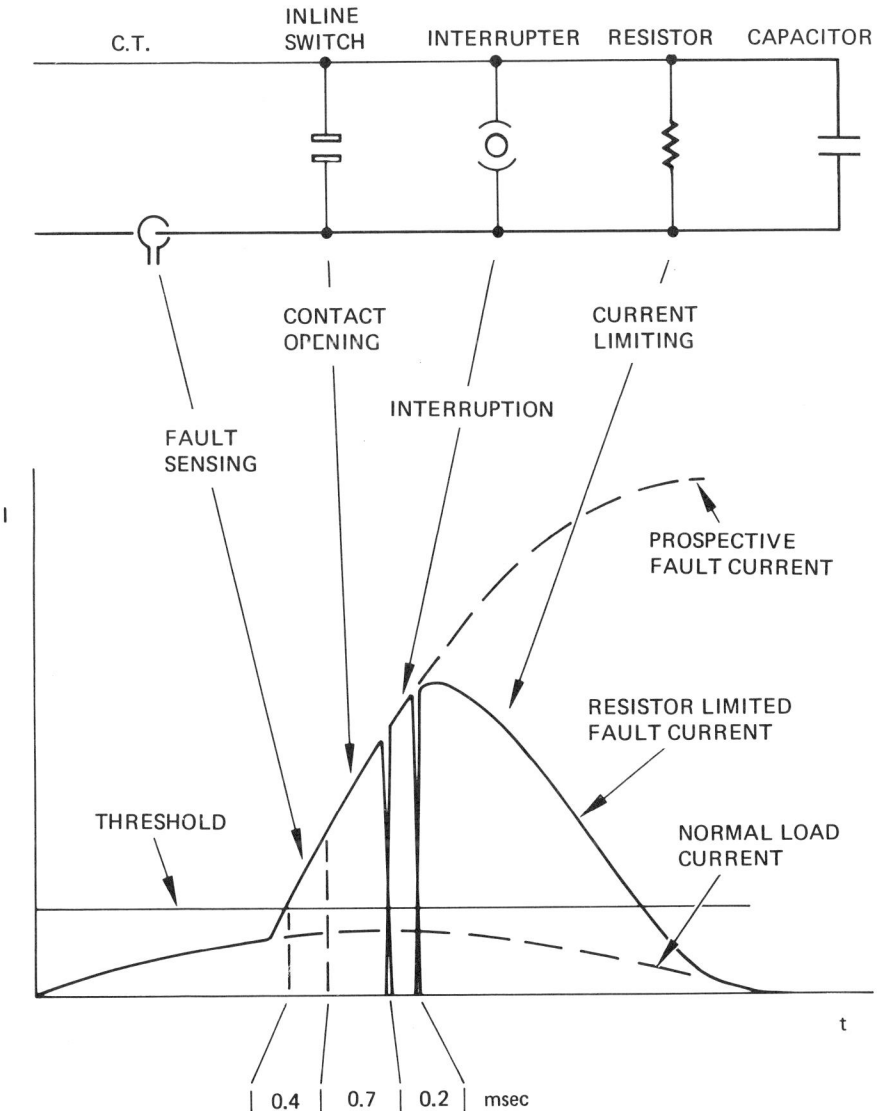

Fig. 9. The current through the FCL is commutated from the contacts of the mechanical switch to the interrupter tube and finally to the resistor to complete the current-limiting sequence.

across each interrupter and each in-line switch contact gap. Figure 10 is a photograph of the interior of the switch module showing the various switching components.

Refer to (7) for details of the in-line switch and interrupter design, construction, and operation. Both devices are actuated by capacitor banks which are maintained charged to 3 kV at all times. One redundant or recycle operation is available from the initial charge — recharge time is 10 sec. The in-line switch consumes a modest amount of SF_6 at each firing. The crossed field interrupters operate from a factory sealed helium reservoir.

The resistors that were designed and provided by American Electric Power (Figure 11), consist of series/parallel stacks of silicon carbide discs. The nonlinear resistor characteristic (Figure 12) of this material is required to accommodate the bounary conditions imposed by the maximum insertion voltage at current interruption (140 kV determined by the interrupters) plus the impedance required to limit the current to 700 A rms after the resistors are inserted. A typical resistor insertion sequence is shown in Figure 13. The energy absorbed by the resistor while in the line is the sum of that in the insertion spike plus that generated by the follow-on current which flows until a breaker clears the fault. The nonlinearity of the resistance material with both current and temperature makes it impossible to calculate the absorbed energy in closed form. Computer calculations indicate that for the resistors used here that have a cold (20°C) resistance of 22 ohms per phase at 6.5 kA (Figure 12) there will be 1.4 MJ absorbed per phase (two resistors, see Figure 9) during a 6.5 kA insertion spike and an average of 1.5 MJ per cycle until the breaker opens. During a 5 cycle insertion the resistor temperature will increase 20°C. Maximum operating temperature for the blocks is 100°C. The resistor assemblies shown in Figure 11 are forced-air cooled.

The power circuits for each phase sit on separate platforms isolated from ground by post insulators as shown in Figure 14. There is one control/command module located under the center platform which controls all three units as described below.

D. Communication and Command System

The principal sensing, communication, and command components are shown in Figure 15. The line current in each phase is monitored by a bushing current transformer and

MC12022

Fig. 10. Interior view of switch module showing in-line switch (foreground), one of the crossed-field interrupters (right), and the capacitor pulsers which actuate the in-line switch and interrupter (bottom).

Fig. 11. Interior of one resistor module showing series/parallel stacks of silicon carbide resistor discs.

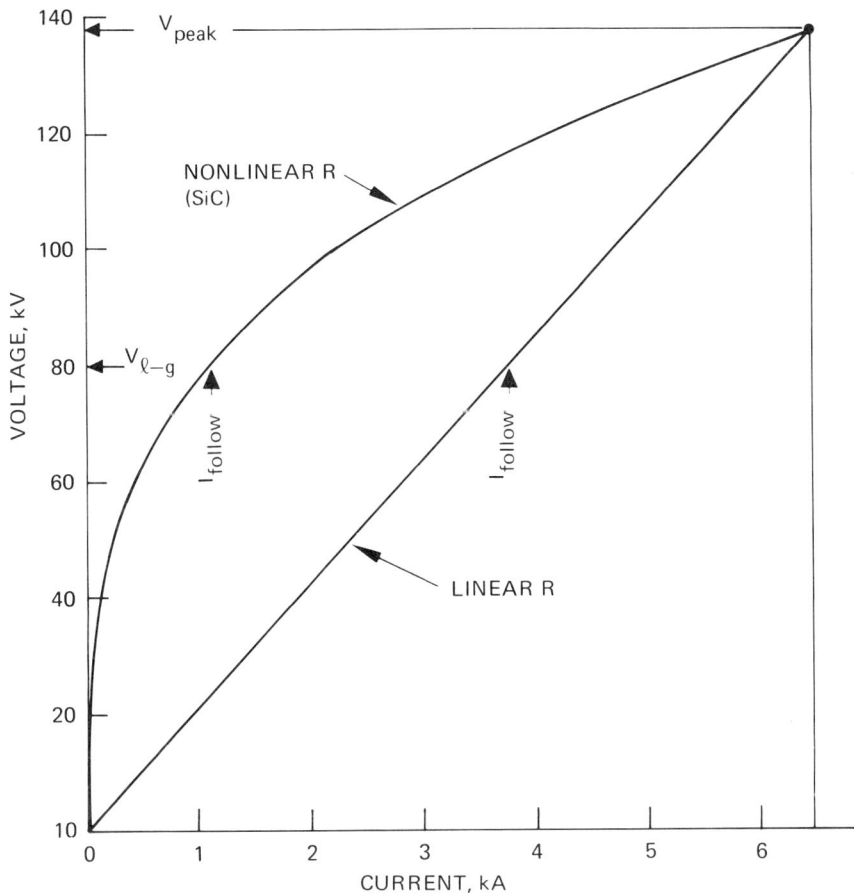

Fig. 12. Follow-on currents for nonlinear and linear resistors after resistor insertion at 6.5 kA when transient insertion voltage is limited to 140 kV.

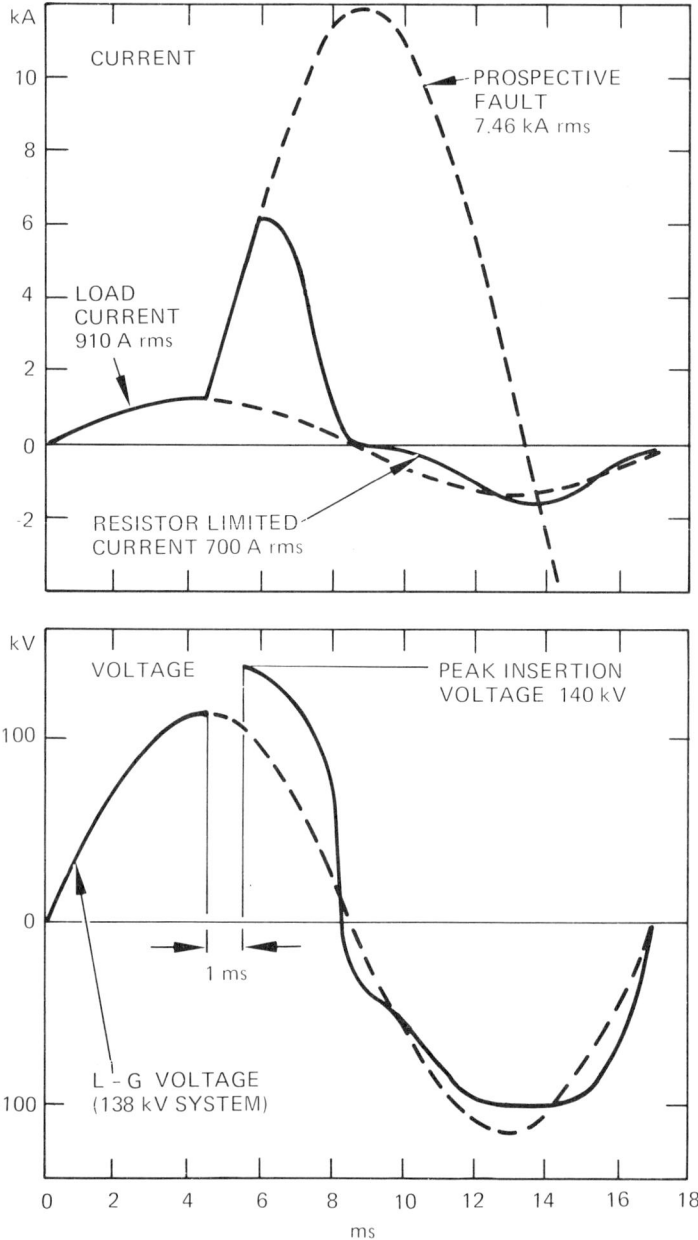

Fig. 13. Calculated transients expected during the resistor insertion sequence for Hughes FCL scheduled for delivery to American Electric Power Corporation.

FAULT CURRENT LIMITERS 159

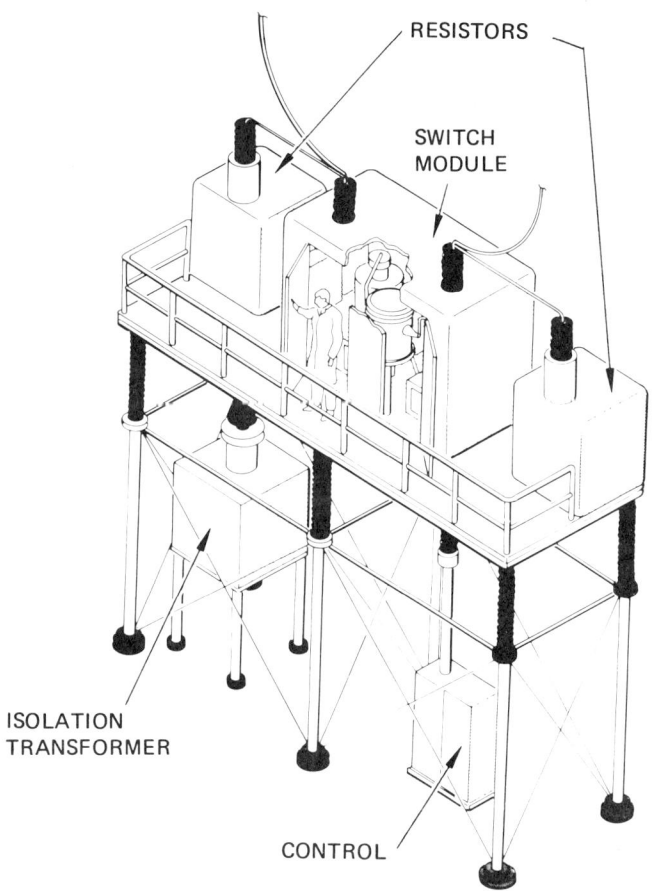

Fig. 14. One phase of 138 kV FCL installation. A single system controls all three phases. Platform is 23 ft x 8 ft.

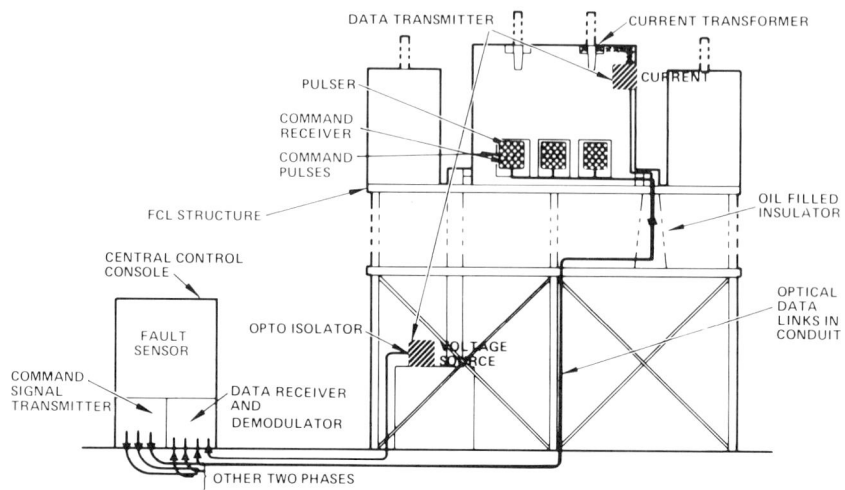

Fig. 15. FCL control system illustrating control and command signals transmitted across the high voltage interface via optical data links to a central console which controls all three phases.

transmitted to the ground control module via an optical data link or "light link." The glass fibers that carry the information in these data links are well suited for the purpose since they are completely dielectric and hence neither compromise the high voltage integrity of the system nor suffer signal degradation in the harsh EMI environment of the switchyard. As described in Section V-B, a 1 kA line current is seen by the fault sensor as a 1 V signal. A voltage signal is derived from the isolation transformer bushing to provide a phase reference for the simulated fault current. With these inputs the fault sensor is able to analyze transients and correctly command the FCL.

The command system that activates the switch module is designed with several "fail safe" features. Commands are sent from ground to the switch module via optical data links that are inherently immune to noise and EMI. The command is sent to the platform over three parallel links via a two bit command word plus a parity check bit and no action is initiated until two consecutive instructions are received verifying the same command. Thus, control of the operating cycle for each phase is given to that switch by the ground control. Finally, a test signal is continuously circulated around the loop formed by the downcoming current link and the upgoing command links. If this test signal is not verified, the entire FCL is deactivated and a "not ready" signal is displayed in the control house.

The system is configured so that the fault sensor can be tested with an internal fault simulator, the command sequence can be tested with the last trigger pulse deactivated (so the switch does not open) or the entire FCL can be manually tripped, all while the system is online.

The reclosure logic and timing may be configured in a variety of ways and must be carefully integrated with the system relaying so that an unstable situation does not occur for evolving faults or a fault that does not clear.

The system also includes a comprehensive system of sensors, logic, and interlocks to ensure that each of the components is properly armed and ready before a switching action is initiated. Since these are not fundamental to the operation of the FCL, they will not be described in detail here.

E. System Test and Evaluation

1. Fault Sensor and Command

A prototype fault sensor has been fabricated and tested first at HAC (see results, Table 2), later on the AEP analog simulator, and finally in the field at the Muskingum River site where the first HAC FCL will be installed. To conduct these field tests a monitoring station was installed at the location shown in Figure 7. Essentially this installation (Figure 16) duplicates the current transformers and light links of the final FCL installation. The line current signals were fed both to the fault sensor as well as to high speed recording equipment provided by American Electric Power. The first series of field tests consisted of a lengthy series of switching operations typical of those employed during normal yard operation. The FCL control recognized these as non-faults and did not operate. Next a series of line-to-ground faults were staged at varying distances along the Natrium line. As expected, the FCL control rejected the far faults as too small to require operation but did trigger on the close-in fault which exceeded its critical value (set at 1.25 kA instead of 5 kA for these tests). Figure 17 shows the response of the fault sensor of this test in which the threshold crossing and operate command are indicated, 400 μsec apart.

Fig. 16. Sensor pods containing current transformers and optical data transmitters used to gather data on system transients and evaluate the FCL controls in the field. Installation is in the Muskingum Power Switchyard of the Ohio Power Company.

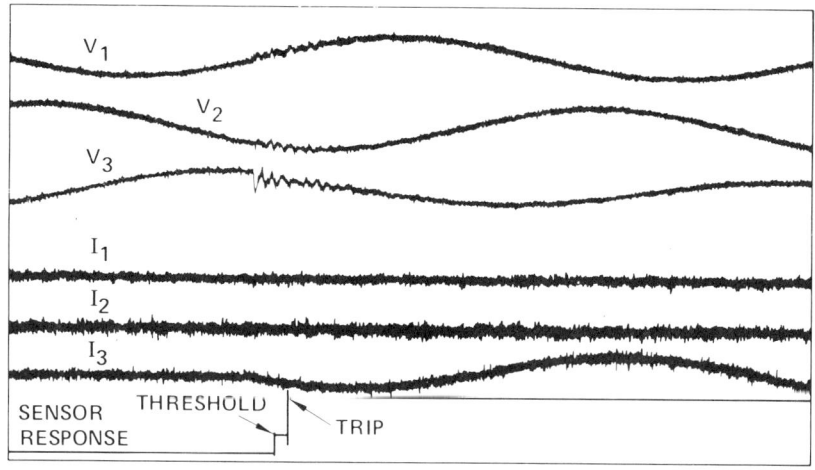

Fig. 17. Test record which illustrates the FCL control correctly interpreting the presence of a line-to-ground fault during a stated test.

2. Switch Modules

Construction of the structural support platforms and the three-switch modules has been completed. Switch module No. 1 has been integrated with both the deliverable fault sensor and the command unit as well as with the factory test facility. Preliminary current commutation tests are presently in progress.

VI. SUMMARY/CONCLUSIONS

The need for fault current limiters is becoming increasingly apparent to those who plan, design, or operate electric utility systems. Enough FCLs are in operation world-wide to indicate technical feasibility. Interest in the subject is high and a considerable body of literature is developing.

Technically, the challenges appear to be in the areas of sensing, speed of operation, repeatability, energy absorption/dissipation, and reliability. There are a number of development projects underway at this writing, working toward the solution of these technical challenges. One further technical area that has received little attention so far is the study of system interaction with FCLs. The effects of their application on system stability, protective relaying, and reliability have not been widely reported. The possible use of FCLs in long, high voltage transmission lines to protect series capacitors and promote transient stability should be studied. Application of FCLs specifically to reduce or eliminate through fault failures of expensive power transformers must be investigated. As interest in the subject spreads, new possibilities will come to light which are not apparent now.

As in all engineering endeavor, technology is not an end in itself. Each new FCL design must compete with all the alternatives in each application. A partial list of alternatives might include heavy duty breakers, higher transformer impedance, reactors, and/or system splitting. Each potential application will require thorough costing of each alternative and a rational decision based on all the costs. It is the authors' opinion that as development of FCL devices progresses, costs will come down to the point where they will compete favorably in more applications as time goes on.

VII. REFERENCES

1. C.A. Falcone, J.E. Beehler, W.E. Mekolites, and J. Grzan, "Current Limiting Device - A Utilities Need," IEEE Trans., PAS-93 (Nov/Dec) (1974), 1768-1775.

2. C.A. Falcone, "Current Limiting - An Alternative Approach to Fault Interrupting," Winter Pwr Mtg., Penn. Electric Assoc. Relay Com., Pittsburgh, Pa, Feb. 1974.

3. Various Authors, Symposium on Current Limiting Devices, IEEE/PES 1974, IEEE No. 75CH01037-1PWER.

4. Various Authors, Symposium on New Concepts in Fault Current Limiters and Power Circuit Breakers, 1976, EPRI No. EL-276-SR, April 1977.

5. T.F. Gallen, "Protective Relaying Implications of Fault Current Limiters," Symposium on New Concepts in Fault Current Limiters and Power Circuit Breakers, Buffalo, 1976, EPRI No. EL-276-SR.

6. "Development of Fault Current Limiters for Electric Power Systems," Final Report on Project RP281-1, prepared by ITE Imperial Corporation (Subsidiary of Gould, Inc.), Colmar, PA, March 1976, EPRI TD-130.

FAULT CURRENT LIMITERS

7. W. Knauer, "Progress Report on a Switched Resistor Fault Current Limiter," Symposium on New Concepts in Fault Current Limiters and Power Circuit Breakers, Buffalo, 1976, EPRI No. EL-276-SR.

8. I. Lee, et al., "An Ultrafast Fault Sensor for Fault Current Limiters," to be presented at IEEE Summer Power Meeting, 1978.

9. B. Kalkner, "Short-Circuit Current Limiter for Coupled High Power Systems," CIGRE Session 1966, Paper No. 301.

10. R. Banks and H.L. Thanawala, "Short Circuit Limiting Coupling as AC System Interconnector," IEEE Conference Publication No. 107, November 1973.

11. W.J. Bellerby, W.A. Mittlestadt, et al., "Bonneville Power Administration 1400 mw Braking Resistor," presented at the IEEE PES Summer Meeting 1974 (T74 433-9).

12. A.B.R. Kumar and E.F. Richards, "A Solid Core Inductor in Lieu of the Braking Resistor to Improve the Transient Stability of Power Systems," presented at 1974 Midwest Power Symposium, University of Missouri, Rolla, October 21-22, 1974.

13. T. Keders and A.A. Leibold, "A Current Limiting Device for Service Voltages up to 34.5 kV," IEEE-PES Summer Power Meeting, Portland, OR, July 1976, IEEE A76-436-6.

14. Calor-Emag Leaflet No. 1196/6E.

15. "Superconducting Fault Current Limiter," Final report on project RP328 prepared by Argonne National Laboratory, December 1976, EPRI EL-329.

16. T. Hirayama, et al., "Developments of AC Type System Interconnectors," IEEE Transactions on PA&S May/June 1972, pp 1085-1091, 71 TP 609-PWR.

17. S. Welk and F. Plonski, "Fault Sensing for a Switched Resistor Fault Current Limiter," the Fourth Annual Western Protective Relay Conference, October 1977, Spokane, WA.

18. "Investigation of Feasibility of Vacuum-Arc Fault Current Limiting Device," Final report on project RP476, State University of N.Y. at Buffalo, October 1977, EPRI EL538.

19. "Development of a Current Limiter Using Vacuum Arc Current Commutation," Final report on project RP564-1, Westinghouse Electric Corporation, Pittsburgh, PA, EPRI EL393, March 1977.

20. "Controlled Impedance Short Circuit Limiter," Interim report on project RP654, Westinghouse Electric Corp., Pittsburgh, PA, EPRI EL537, September 1977.

Power Control System and Protection

HYBRID SIMULATION OF ELECTRIC POWER SYSTEMS

Dr. Lewis N. Walker

University of Missouri - Columbia

I. INTRODUCTION

Due to speed, economy, and safety features, simulation techniques have been widely used by analysts to gather information about problems. In many practical situations, the mathematical equations describing the problem are well known, and development of a mathematical model is straightforward. Three main simulation classifications are digital simulation, analog simulation, and hybrid simulation (mixing both digital and analog simulation techniques).

Digital simulation represents parameters and problem variables as discrete quantities, and produces approximate solutions to the mathematical equations of the model. A typical digital simulation is run using a digital computer. The resolution and accuracy of the digital computation are limited by the number of bits used to represent a parameter and how fast it is sampled. Thus the resolution can be made as high as desired, subject to economic and state-of-the-art constraints.

Classical analog simulation is accomplished by setting up an analog computer to solve the model equations for the real-world system of interest. The variables of the system under study are represented by analogous physical quantities that can vary in a continuous manner. In the electronic analog computer, electrical voltages are used to represent the variables of the modeled system, and the solution of differential equations is particularly well suited to the analog computer (1). Usually problems are set up on an analog computer by patching the proper computation components together through the use of a patch panel so that the resulting circuit is a simulation of the problem. Computation components can add, subtract, integrate, differentiate, multiply, divide, or generate non-linear functions. The solution rate of the problem can be scaled to be either faster or slower than real-time.

Hybrid simulation combines the advantages of both the

analog and the digital techniques. It can perform faster than real-time simulation using analog computation elements that yield a continuous-time solution. Higher resolution of certain parameters can be achieved by using digital techniques. Furthermore, simulation problem change-overs can be extremely fast, especially those that do not require scaling changes.

II. ADVANCED HYBRID COMPUTATION

Advanced hybrid computing has become increasingly important in recent years due to cost and speed advantages over purely analog or purely digital computers (2-6). A recent special Symposium on Advanced Hybrid Computer Systems (AHCS) held in San Francisco in mid-1975 focused on the latest development in the highly reliable, low-cost, solid-state switching devices, advanced communication techniques and high-level hybrid programming languages. All these provide the AHCS with automatic patching and scaling, remote terminal access for "time-slicing" the system and complete programmability in a high-level language. AHCS will maintain a 30:1 cost and speed advantage over pure-digital systems for solving dynamic problems. A survey conducted by the U.S. Army Materiel Command in 1974 indicated that over 67% of potential industrial users and more than 83% of potential academic users would use an advanced hybrid computing system remotely via a data communications network if available (7).

III. HYBRID SIMULATION OF ELECTRIC POWER SYSTEM MODELS

In recent years hybrid simulation has become an important tool to study the behavior of electric power system components and power system response (8-11).

Hybrid simulation of electric power systems has been considered by a number of investigators. A hybrid load flow computer was developed jointly by the New England Power Exchange (NEPEX) and the Westinghouse Electric Corporation for the control of the NEPEX bulk transmission system (12,13). Electronic Associates Incorporated (EAI), supported by the Edison Electric Institute (EEI), chose to examine the high speed parallel computing capabilities of hybrid computers (14, 15).

For the past five years the University of Missouri - Columbia has been involved in hybrid simulation of electric power systems. During this time two unique simulators have been developed. The first was an overvoltage simulator (OVS) designed to produce the overvoltages that result from transmission line switching (16, 17). The OVS is a three phase (plus ground) distributed parameter simulation with a digital controller to control switching breaker pole span and metering.

The second simulator was the faster-than-real-time long term dynamic hybrid simulator developed under the EPRI RP908 -1 project at the University of Missouri - Columbia (18, 9). The long term dynamics hybrid simulator was modified in the RP908-2 project to be valid for the short term dynamics and to improve the set up and operation (19). Since the EPRI hybrid simulator is more general primary attention will be devoted to the long term dynamics simulator.

The large scale hybrid computer, simulates a 16-generator and 68-bus electric power system. The goals were to demonstrate the advanced hybrid computation techniques in faster than real-time simulation and to allow a continuous simulation run out to 20 minutes (real-time) of system response. The power system model is an ac, single phase, variable frequency, 20 times faster than real-time simulation implemented in electronic analog circuitry. A 16-bit TI 980A digital minicomputer is used for monitoring and control of the analog system. A special purpose digital-analog interface controller was designed and built to achieve the desired data collection and control functions.

Figure 1 shows a picture of the hybrid simulator and a block diagram of the hybrid simulator is shown in Figure 2. The entire power system model is simulated in analog circuitry, except for the slow nonlinear boiler characteristics of a generation unit, which is simulated digitally. Type 741 integrated-circuit operational amplifiers and type 532 four-quadrant multipliers were used throughout the analog simulation. All the analog models are assembled onto printed circuit boards and are plugged into card racks.

The special purpose digital controller, which includes the analog-digital data transfer system, consists of a Texas Instruments model 980A digital computer, a disc, a dual Mohawk cartridge tape unit with a custom designed controller, a card reader, a silent-operation teletypewriter, and a cathode-ray-tube display. The custom designed interface hardware between the digital computer and the peripherals contains analog-to-digital converters, digital-to-analog converters, electronic switches, and a direct-memory-access-channel (DMAC) data transfer system.

Power System Hybrid Simulator

Figure 1

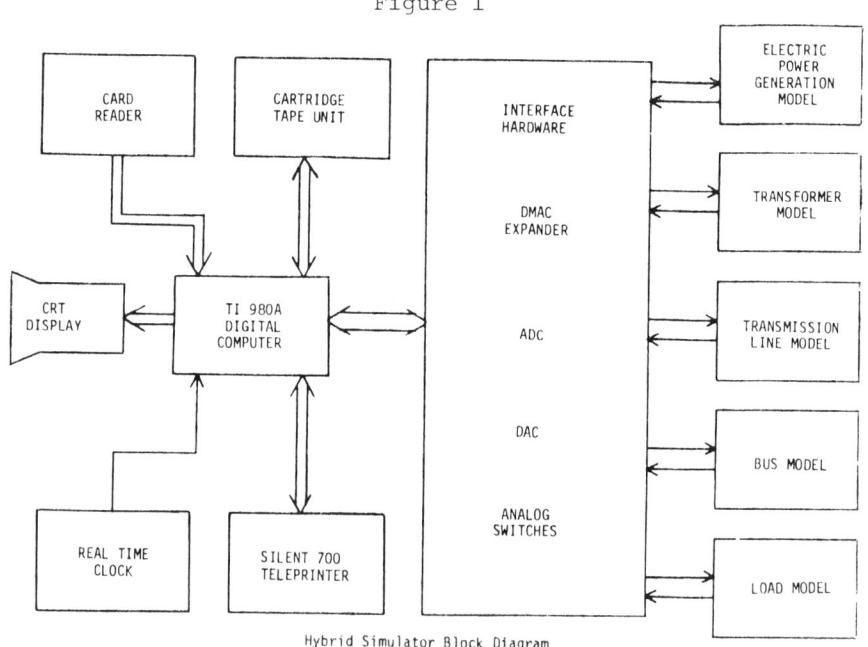

Hybrid Simulator Block Diagram

Figure 2

HYBRID SIMULATION OF ELECTRIC POWER SYSTEMS

The hybrid simulator is housed in a five-bay 19-inch cabinet. Two 1632-point patch panels are used to interconnect the electric power system configurations under study. All inputs and outputs of the analog models are brought to the patch panels using coaxial cables. Each 1632-point (48 x 34) patch panel has a dimension of 13" x 9" in which more than 1,000 points are used.

There are 230 8" x 10½" printed circuit boards assembled in this system. A total of more than 4,000 integrated-circuit operational amplifiers were used in the analog computation, and more than 3,000 digital integrated circuits were used in the digital-interface circuits, monitoring circuits, and control circuits.

In the hybrid simulation of electric power systems, analog simulation was used for the basic elements of the power system; namely, generation unit, transformer, transmission line, bus, load, etc. All these five major electric power system components are implemented under the EPRI RP908-1 and -2 projects. A mathematical representation of different models was first formulated to include all functions that have significant effects on the dynamics of the power system. An analog circuit was derived to represent the mathematical equations and implemented using integrated-circuit operational amplifiers and multipliers. Since some of the very long time constants for the generation unit models can be better simulated digitally, the boiler control, boiler pressure of the generation units and the automatic generation control were simulated digitally.

An AC variable frequency model was chosen because 1) it minimizes the effect of DC drift in the amplifiers, 2) accurate integration to compute frequencies and angles is not necessary to maintain computational stability, 3) a direct representation of the system elements in action is possible including bus and phase angles, 4) the model responses can be more directly compared with system behavior, and 5) the variation of impedance with frequency is included.

A special purpose digital controller was developed and the functions performed can be summarized as follows: 1) initiate and control all operations performed by the simulator, 2) process all data for the simulator, 3) respond to interrupt conditions from the model, 4) provide documentation logging of all operations performed by the simulator, 5) simulate the very slow speed of the non-linear boiler functions, and 6) perform various monitoring and control functions.

Long-term monitoring and control are set up at a typical scan rate of two seconds (real time) for the power system. Voltages, currents, phase angles and bus angles are monitored at every bus. High speed sampling channels measure phase angles which allow fast samplings of swing angles. The typical sampling interval is two seconds (real time) for digital simulation on the boiler dynamics, but it can be increased if necessary.

With this hybrid approach, the digital storage and computing requirements have been greatly reduced as compared with the all digital simulation technique. Network parameters do not have to be stored in the digital computer since they are represented in the analog computation models. The total computing time for a problem solution is also reduced because differential equations are solved by the analog circuitry. Also, the settling time for the analog network to reach a steady-state solution which satisfies boundary conditions is decreased.

Recent developments of inexpensive, analog integrated circuit components have allowed the special purpose analog computer to successfully compete in the solutions of certain classes of problems. Moreover, the development of inexpensive hybrid components, multipliers, digital to analog converters and analog switches, have made the solutions to this problem feasible through the use of hybrid computers (5).

The following are characteristics of the type of problem which is amenable to analog computation:

High speed solutions: The speed of a digital solution is limited by the cycle time of the digital computer and the number of steps required for the solution. Even with fast computer cycle times the speed of an analog solution often cannot be attained.

Simultaneous solutions (simultaneous differential equations): A digital computer normally obtains many simultaneous solutions by "time sharing" the processor with each solution. Every solution must use the intermediate results of every other solution during computation. Iterative techniques which obtain one solution at a time may also be used. Either of these methods may be satisfactory if computation time is not critical.

An analog computer can solve this type of problem by a simultaneous solution of all problems with parallel, interconnected analog hardware. In order for a digital computer to work in the same manner, parallel processors with shared memory would be required. The present state-of-the-art in parallel processors (both hardware and software) is rather limited and those available at present are very expensive.

When these two requirements for high speed and many simultaneous solutions are combined, the need for analog computation becomes critical. In some cases, a high speed solution is not only desirable in order to reduce computation costs, but is required to make the solution meaningful (8).

In addition to the analog hardware, a hybrid system contains digital portions to implement selected functions. These functions best implemented in a digital fashion are 1) solution of the nonlinear portions of the problem; an example is the simulation of boiler characteristics (requiring the use of steam tables) of an electrical power generation unit, 2) modification of the problem (or parameters of the problem) as a result of the solution in progress; an example would be the opening of an electrical circuit breaker when the current in that simulated breaker exceeds its preset limit, 3) pre-programmed modification of the problem while the solution is in progress; an example would be to remove a transmission line of a power system simulator in order to simulate an open line, and 4) monitor the analog system and store data for later analysis; a large scale analog system may contain hundreds of points of interest which could most effectively be analyzed after completion of the experiment.

IV GENERATION UNIT MODEL

The generation unit is the heart of the electric power system, which includes a very detailed representation of various components. Both analog and digital simulation techniques are used. The division was chosen in such a way that the generation unit can function without the digital portion (20,21). With only the analog portion operating, the Load Reference Motor (LRM) position, Throttle Pressure (PTH), and Delta Control Value Position (DCVP) are held constant. A complete block diagram is shown in Figure 3.

A. Analog Simulation of Generation Unit Model

The analog portion of the generation unit model is divided into four parts. Each part is assembled on a printed circuit board as follows:

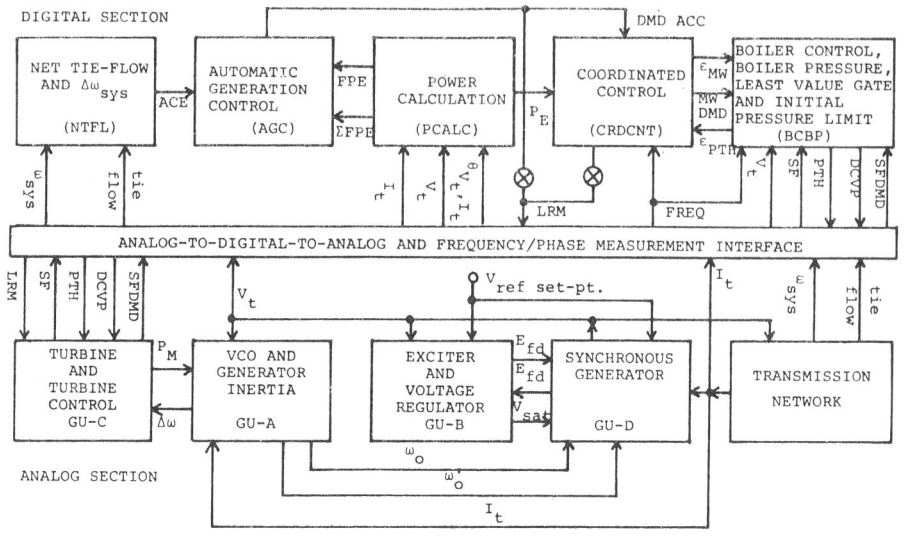

Generation Unit Block Diagram

Figure 3

1. GU-A. Voltage controlled oscillator and generator inertia

The heart of the generation unit model is the voltage controlled oscillator (VCO). The outputs of the oscillator (ω_0 and ω_0') are constant magnitude, variable frequency sine waves, phase-shifted $90°$ with respect to each other. These signals are used to synchronously detect and modulate the d- and q-axis components of current and voltage. The terminal voltage (V_t) and the terminal current (I_t) are used to calculate the real electrical power output, which is then compared with the mechanical power (P_M) to calculate the frequency error ($\Delta\omega$).

2. GU-B. Voltage regulator and exciter.

The GU-B printed circuit board models the generator voltage regulator and exciter. The voltage regulator rectifies the terminal voltage (V_t) and calculates a signal to feed into the exciter model. The exciter model calculates the field voltage (E_{fd}). The magnitude of V_t is set by the reference set point ($V_{REF\ SET\ PT.}$). The output of the terminal voltage saturation detector (V_{SAT}) connects to the synchronous generator board.

3. GU-C. Turbine and turbine control.

The third generator PC board contains the turbine and some turbine control. A steam flow demand (SFDMD) signal is calculated from $\Delta\omega$ and the LRM setting. The SFDMD signal connects to the digital model, which calculates a delta control valve position (DCVP) signal. The DCVP signal is added to SFDMD and then a steam flow (SF) is calculated. The throttle pressure (PTH) input is also used to calculate SF.

4. GU-D. Synchronous generator.

The synchronous generator part of the generation unit model is included on the fourth PC board. It contains circuitry to calculate V_t by synchronously detecting the d- and q-axis currents, calculating the d- and q-axis voltages, and summing the voltages to produce V_t.

The generation unit model supplies a voltage to the transmission network and receives a feedback current. The tie-line power flows and the system frequency are obtained from the transmission network.

B. Digital Simulation of the Generation Unit Model

Five separate programs make up the digital portion of the generation unit model as follows:
1. Net tie-flow and $\Delta\omega_{sys}$ (NTFL)
2. Automatic generation control (AGC)
3. Power calculation (PCALC)
4. Coordinated control (CTDCNT)
5. Boiler control, boiler pressure, least value gate, and initial pressure limiter (BCBP)

The NTFL program calculates an area control error (ACE) signal from the system frequency and the tie flows. The ACE connects to the AGC program.

The PCALC program receives the magnitude of the terminal voltage and current and the phase between them. From this information, PCALC calculates the real electrical power (which connects to CRDCNT), a filtered electrical power (FPE), and the sum of all the FPE's (ΣFPE) in a particular AGC area.

The AGC program contains economic dispatching and

frequency control. It can be used in two modes; one is with
CRDCNT and the other is without CRDCNT. With CRDCNT the output of AGC is a demand accumulator output, which is input to
CRDCNT. Without CRDCNT, the output of AGC is the LRM.

The CRDCNT program calculates the LRM when it is used.
It also calculates the MW error and the MW demand signals
that connect to the BCBP program.

The main functions of the BCBP program are to calculate
the throttle pressure (PTH) and the DCVP signal. It also
calculates a PTH error signal, which connects to CRDCNT. The
frequency and voltage inputs are used in calculating the
auxiliary outputs.

V. COMPUTERIZED PATCHING (22)

The feasibility of computerized patching and parameter
setting were studied. The detailed designation includes the
general background, a revision of the analog models for modular compatability removing line and bus differentiation
circuits and a description of switch minimization techniques.
Reference 22 gives complete details concerning the computerized patching study performed by UMC.

Two minimization techniques that minimize the number of
switches required for a COMPAT system were studied: the
restricted-connection technique and the modularization technique. The restricted-connection technique takes advantage
of the restrictions of the interconnectivity between different
power-system components. The rules for this technique are
identified; briefly stated: connections are allowed for any
power-system components to coupling networks. Therefore a
saving in the total number of switches required for a COMPAT
system has resulted. A second level of restricted connections
is the so called priority, restricted connection. At this
level a priority is set up for the allowable connections of
different power-system components, which results in further
savings in the number of required switches.

The modularization techniques takes advantage of the
topology of a power system. After the individual modules are
separated from a power system, the intermodule connections are
then made in an intermodule-connection block that consists of
only transmission-line models and switching matrices. The
optimum module size that gives the maximum savings in the
number of required switches is determined. This modularization technique produces a 24:1 savings over the conventional-

connection technique.

The design of a specific COMPAT system--one that includes the COMPAT-system specifications--is presented (22). Figure 4 shows a block diagram of the COMPAT system. A description of the strictly nonblocking, multi-stage, switching matrix is discussed along with an experimental evaluation of a CMOSFET switch-string that can be used in the COMPAT system. A 3-stage switching matrix was chosen for use in the COMPAT system and the optimum choice of the parameters used in the 3-stage switching matrix is derived, based on the criterion of minimizing the number of required switches. A digital data-transfer controller used to control the COMPAT system is described along with a discussion of the operation of the COMPAT system. It is shown that such a COMPAT system suitable for a large-scale PSHS can be implemented.

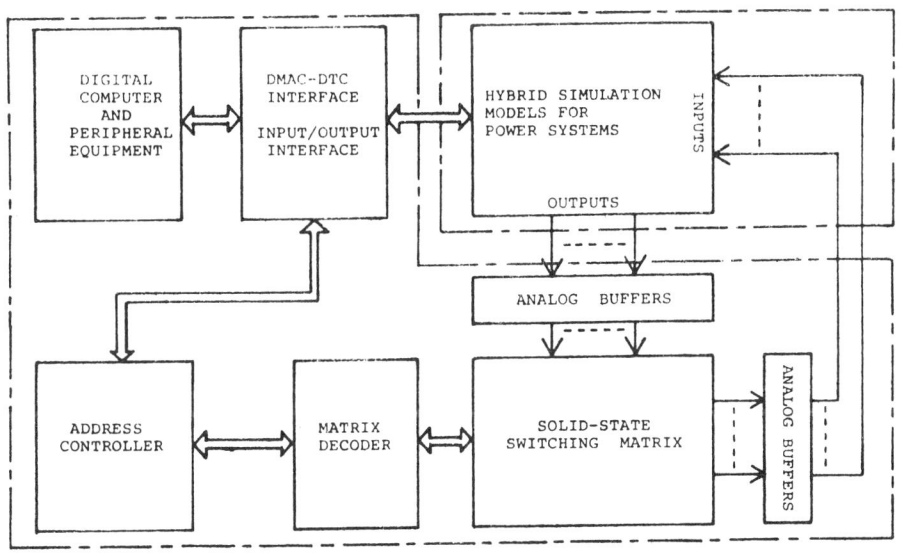

Figure 4
COMPAT System Block Diagram

VI. HYBRID SIMULATOR PERFORMANCE EVALUATION

As a part of the EPRI RP908-2 project the hybrid simulator performance was evaluated. Certain aspects of the hybrid simulator evaluation are presented below.

A. Speed Evaluation

The hybrid simulator was evaluated on several aspects of speed of operation. These aspects include the settling time from a "cold start", the time required to read in data from peripheral devices, initialization time for the digital part of the hybrid simulator models, transient settling time after disturbance, data collection time, and the repeated operation rate.

1. Transient Settling Time After A Disturbance

The settling time of the hybrid simulator following a disturbance is determined by the time constants in the models. Depending on the severity of the disturbance and the definition of steady-state, the settling time can vary over a wide range.

In the present models used with the hybrid simulator, the longest analog time constant is the transient droop washout time (T_r) in the hydro turbine model. Considering this time constant as the worst case time constant should allow all transients to settle. The value used for T_r is 64 seconds in simulation time or 3.2 seconds actual time. Allowing five time constants to reach steady state within 1% would give a settling time of 16 sec.

The settling time could be decreased by reducing the long time constants in the analog model during settling and then increasing them back to the correct value for the simulation.

2. Data Collection and Storage Time

The data collection and storage time depends on the number of data points. Two different methods of data storage are described with each having certain advantages compared to the other. The worst case data collection time is 0.44 ms.

One method of data storage is by real time data logging. With this method the real time program writes the collected data to a peripheral device in real time, i.e., as the data is collected. This requires only enough buffer memory to store one record (or time snapshot) of data. The time required to write one record of data to magnetic tape is about 50 ms plus 0.167 ms times the number of data words. The number of data words is limited by the amount of time allowable between snapshots.

A second method of data storage is to define a large

buffer memory area to store the data as it is collected. Following the end of the simulation run, the data can then be written on a peripheral device. This method allows a much faster data collection rate, but requires a large amount of computer memory. With this method, the data collection rate is essentially limited only by the data collection time and memory to memory transfer time for each data word.

3. Repeated Operation Rate

One of the advantages of the hybrid simulator is that a large number of simulation runs can be made in a relatively short period of time compared to digital transient stability programs.

The necessary steps in a repeated operation mode are:
1. Initialize analog part of hybrid simulator
2. Initialize digital part of hybrid simulator (optional)
3. Run scenario and collect data.

These three steps are then repeated as many times as desired.

The first initialization of the analog part of the simulator requires reading in the analog switch and DAC data. From then on, only the data that is changed during the scenario needs to be reset for the analog initialization. The initialization time is equivalent to the transient settling time discussed in an earlier section.

The digital model initialization is optional depending on the time length of the scenario desired. For very short term scenarios (0 to 10 sec.), the digital model would have negligible effect on the result since it contains long time constants. For longer scenarios, the digital model would be desirable. The first initialization of the digital models may require the worst case time of 138 ms. However, subsequent runs with very similar analog initial conditions would require no initialization of the digital model or a relatively short initialization time.

The scenario run time depends on the hybrid simulator time scaling. The present simulator is scaled to twenty times faster than real time. Thus, a 10 second run would require only 0.5 seconds of actual time.

The repeated operation rate is determined by the sum of the analog initialization time, the digital initialization time and the scenario run time. For a 10 second run, the repeated operation rate could be about 3.6 runs per minute or over 200 runs per hour, assuming worst case settling of a hydro unit. For observed settling time the repeated operation rate could be about 20 runs per minute or 1200 per hour. It is felt that the initialization plus settling time could be cut back to 1.5 sec with no degradation in performance or accuracy, thus producing a total time of two seconds for each run. If further optimization is necessary with faster initialization and settling procedures, it is felt that .5 sec. for initialization and settling could be achieved thus requiring one sec. for each run.

4. High Speed Start-Up and Resynchronization

In order to take full advantage of faster-than-real-time operations, fast initialization of the hybrid simulator is required. One method of synchronization utilizes coherent multimachine system islands. These islands can be initialized directly and quickly with the boiler dynamics held constant. The settling time of a five machine island disturbed by step initialization changes has been tested and is less than 30 seconds (60 Hz base) for most disturbances. Once the islands have been initialized, then the transmission lines connecting the islands together can be switched in.

For a more general rapid synchronization of a system, a synchronization signal is injected through multiplying DAC's into the current input of each generator electrical power calculation circuit. The DAC's can be adjusted to bring the system to previously determined steady state conditions. Then the system currents are switched into the current input of the electrical power calculation circuits, replacing the injected synchronization signals. The inertia and field time constants could also be reduced to further shorten the settling time.

This synchronization scheme was tested and shown to be effective. The transients that occur when switching from the synchronization signals to the system currents are negligible if the initialization is the same as the steady state conditions.

B. Selected Test Results

The hybrid simulator is useful for a wide variety of studies of power systems. Several different types of studies have been performed at UMC utilizing the hybrid simulator. This section summarizes test results taken from the hybrid

simulator to illustrate the flexibility and capability of the hybrid simulator. Further details and information concerning each of the studies are available in the references. This section presents typical results from the following studies:
1. short term dynamics
2. long term dynamics
3. relay sensitivity

Other studies have been performed and a wide range of additional functions could be performed.

1. Short Term Dynamics (21)

The WSCC 9 bus test system was set up on the hybrid simulator. The results of the hybrid simulator are compared with the WSCC system program and the output of the Philadelphia Electric program. Figure 5 shows a one-line diagram of the system and Table 1 shows the system parameters. Figures 6-9 shows the rotor swing curve for a 3φ fault at bus 2 removed after .083 sec. by opening line 4 (GEN 2 230) to 7 (STATION A230). Figures 6 - 7 present swing curves of the rotor of machine 2 with respect to machine 1. Figure 6 shows a comparison between the UMC hybrid, the WSCC digital program and the PE digital program. Figure 7 presents a comparison between UMC hybrid and the WSCC digital. Figure 8 - 9 presents the same information as 6 - 7 for the rotor of machine 3 with respect to machine 1. As can be seen from the plots, the UMC hybrid compares very well with the WSCC digital, and not as well with the PE digital program which is consistant with the differences seen on a one machine infinite bus test system.

TABLE 1

WSCC TEST SYSTEM DATA

Transmission Line Data (100 MVA Base)

Bus	to Bus	R(%)	X(%)	B(%)
2	7	1.00	8.50	17.6
2	8	1.70	9.20	15.8
4	7	3.20	16.10	30.6
4	9	0.85	7.20	14.9
6	8	3.90	17.00	35.8
6	9	1.19	10.08	20.9

Transformer Data (100 MVA Base)

Bus to Bus		R(%)	X(%)	Tap
1	2	0.0	5.76	1.0
3	4	0.0	6.25	1.0
5	6	0.0	5.86	1.0

No saturation effects included

Generator Data (All per unit values are on machine rating base)

Synchronous Rotor Data

Generator 1 and 3 are simple single axis models, while Generator 2 is a two axis representation. No damping or saturation effects are included.

Parameter	Gen. 1	Gen. 2	Gen. 3
rating (MW)	240	180	130
H (pu)	9.85	3.56	2.32
x_d (pu)	X	1.612	X
x_d' (pu)	0.146	0.216	0.236
x_d'' (pu)	X	0.153	X
x_q (pu)	X	1.556	X
x_q'' (pu)	X	0.354	X
T_{do}' (sec)	X	6.0	X
T_{do}'' (sec)	X	0.035	X
T_{qo}' (sec)	X	0.54	X
R_a (pu)	X	0.0002	X

Exciter Data

Generator 1 and 3 are fixed field, while Generator 2 uses the IEEE Type 1 exciter model.

Generator 2 data only

$T_R = 0.06$ sec $K_E = 0.0$

$K_A = 20$ $T_E = 0.314$ sec

$T_A = 0.2$ sec $S_{E_{75max}} = 0.104$

$V_{R_{min}} = -1.167$ $S_{E_{max}} = 0.293$

$V_{R_{max}} = 1.167$ $K_F = 0.063$

$E_{fd_{max}} = 3.984$ pu $T_F = 0.35$ sec

Governor and Turbine Data

No governor or turbine data is required. All three generators are modeled with constant mechanical power.

Load Data

All three loads are constant impedance.

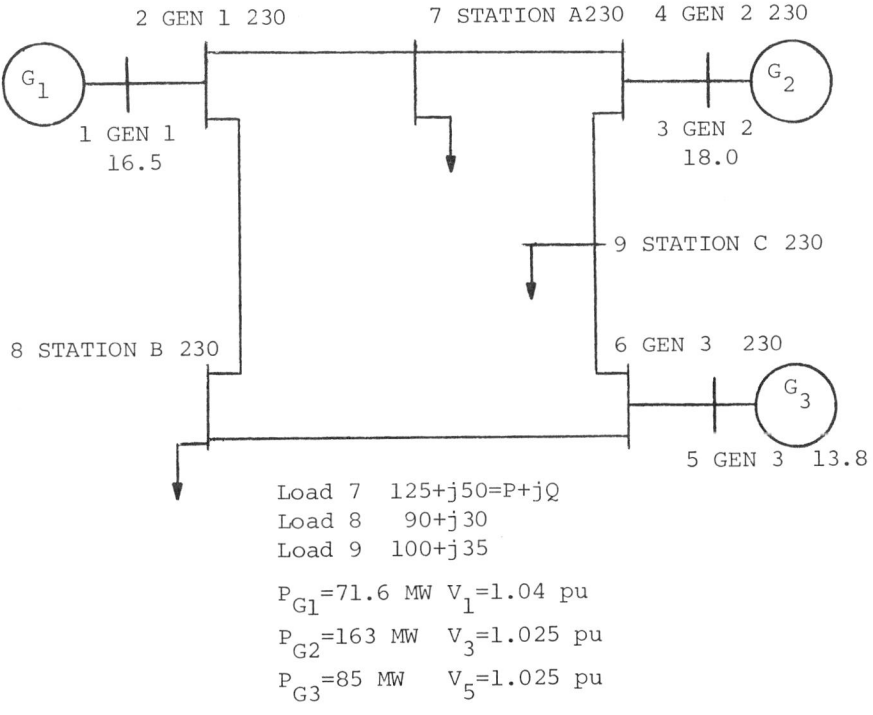

Load 7   125+j50=P+jQ
Load 8   90+j30
Load 9   100+j35

P_{G1}=71.6 MW V_1=1.04 pu
P_{G2}=163 MW V_3=1.025 pu
P_{G3}=85 MW V_5=1.025 pu

FIGURE 5
WSCC One Line Diagram

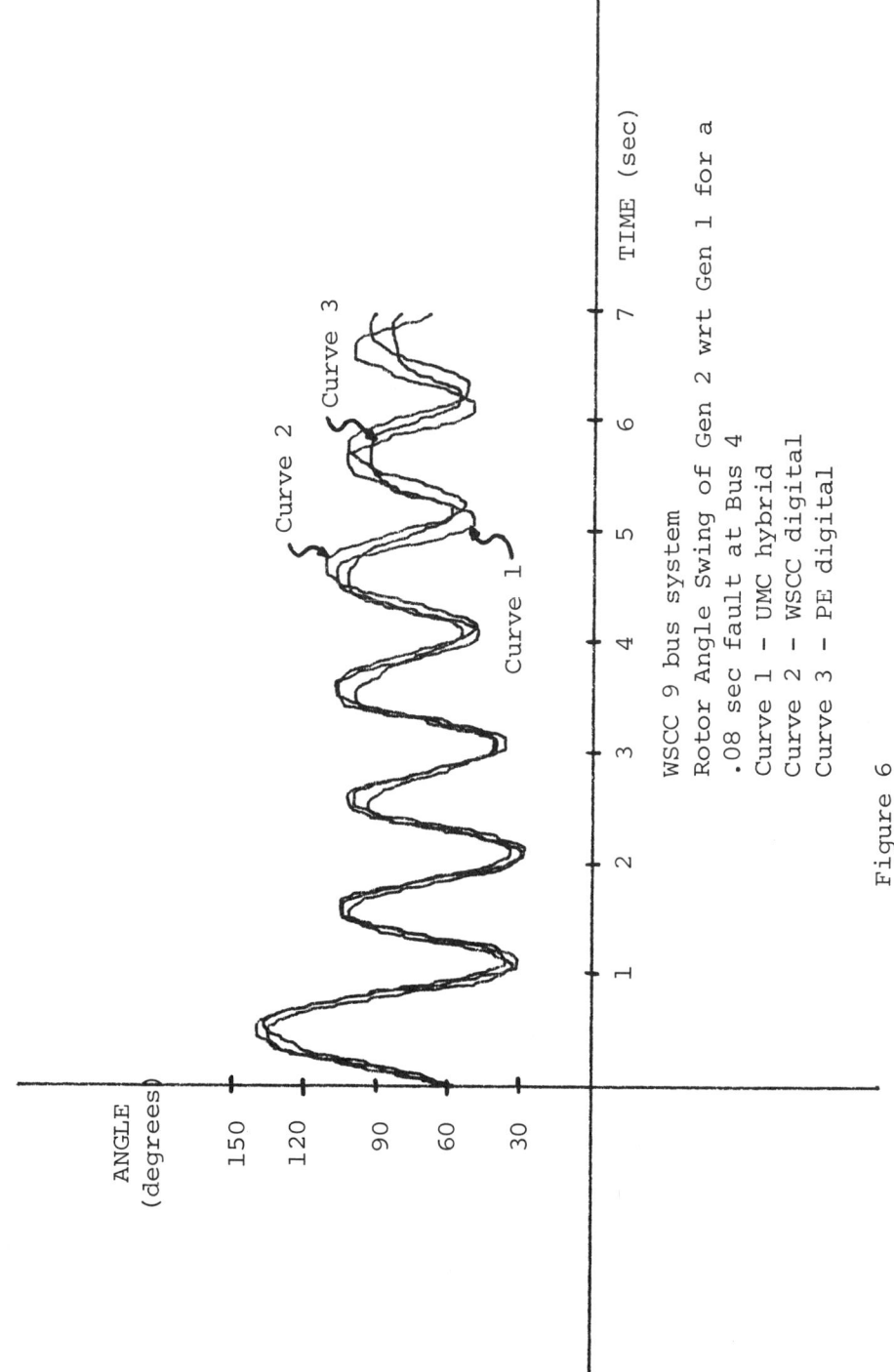

Figure 6

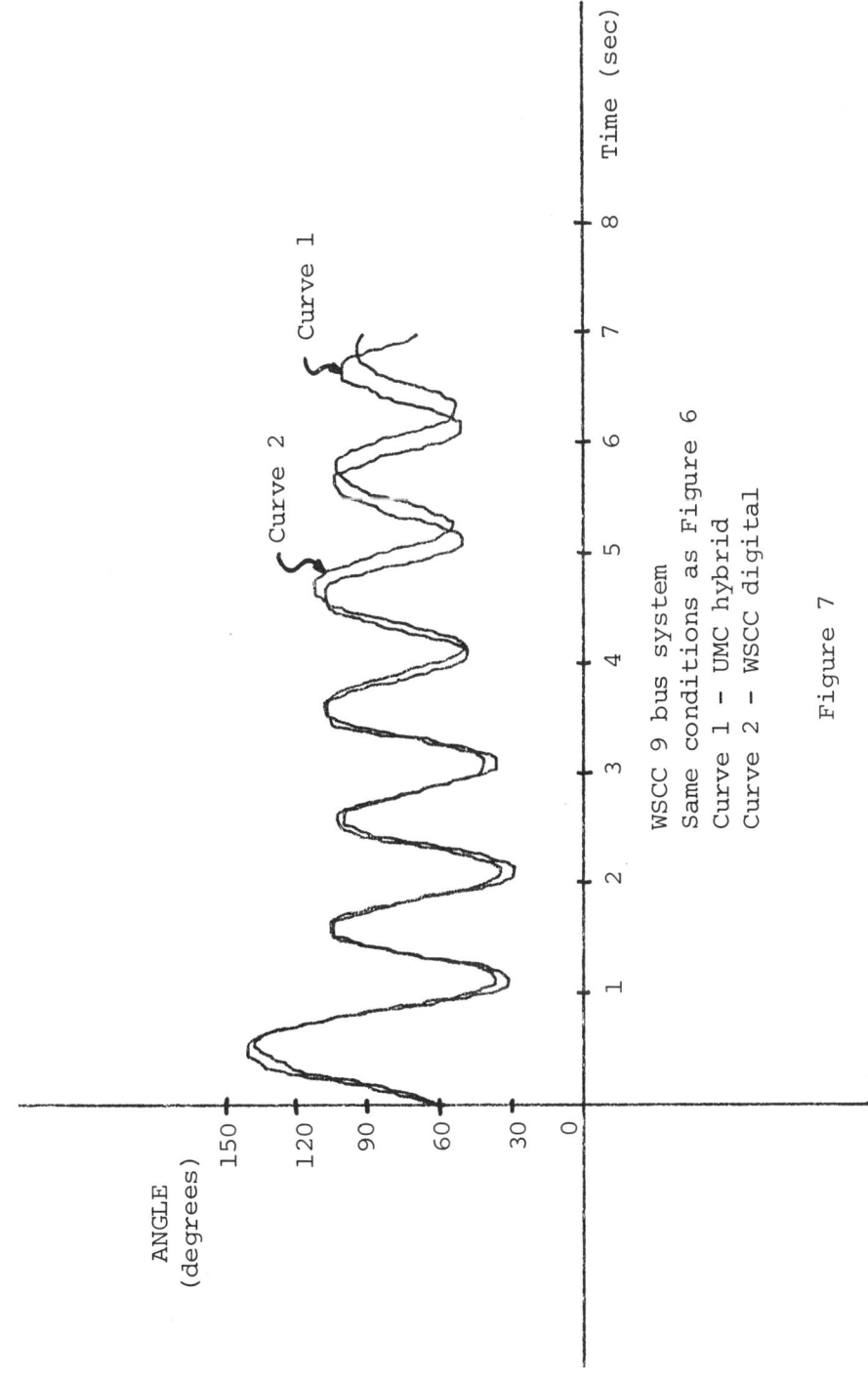

Figure 7

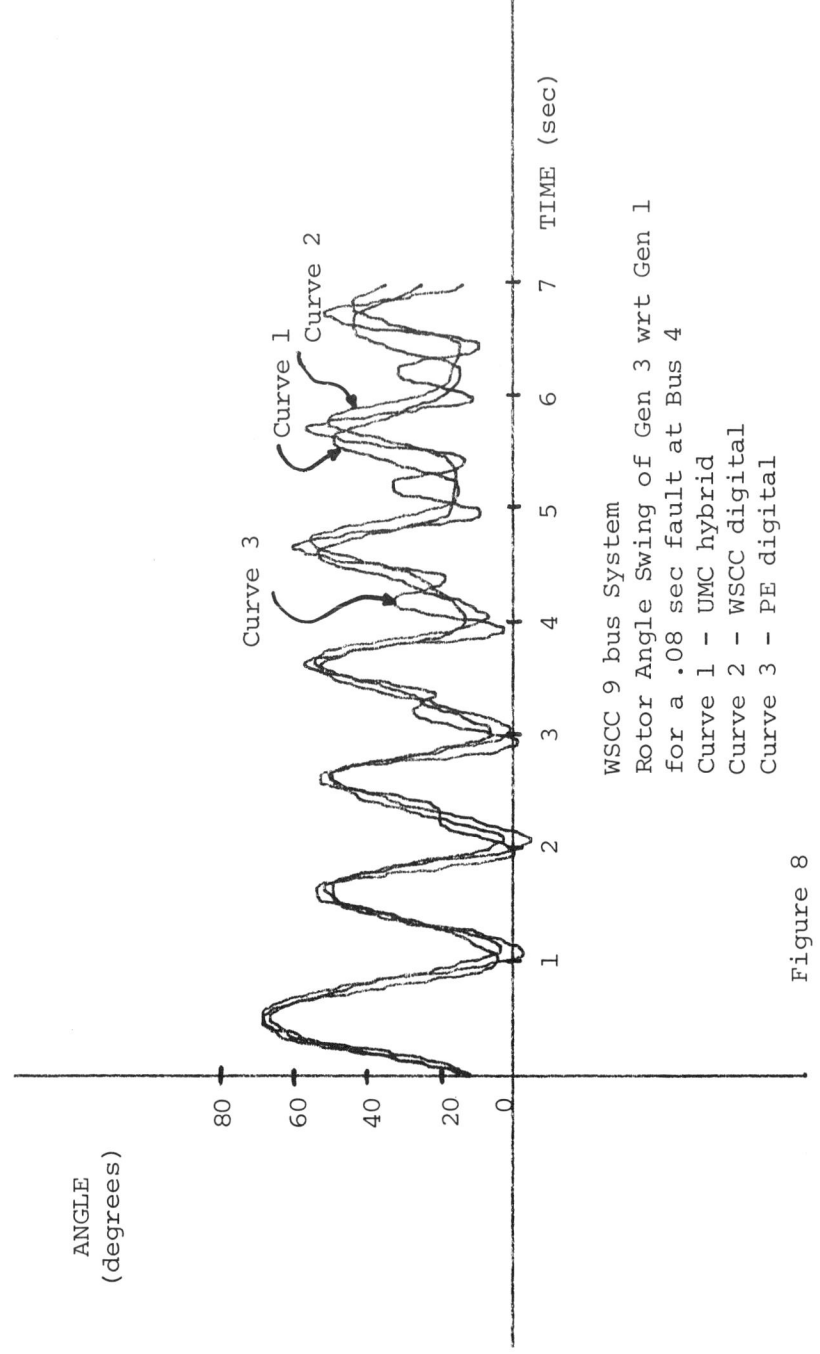

Figure 8

WSCC 9 bus System
Rotor Angle Swing of Gen 3 wrt Gen 1
for a .08 sec fault at Bus 4
Curve 1 - UMC hybrid
Curve 2 - WSCC digital
Curve 3 - PE digital

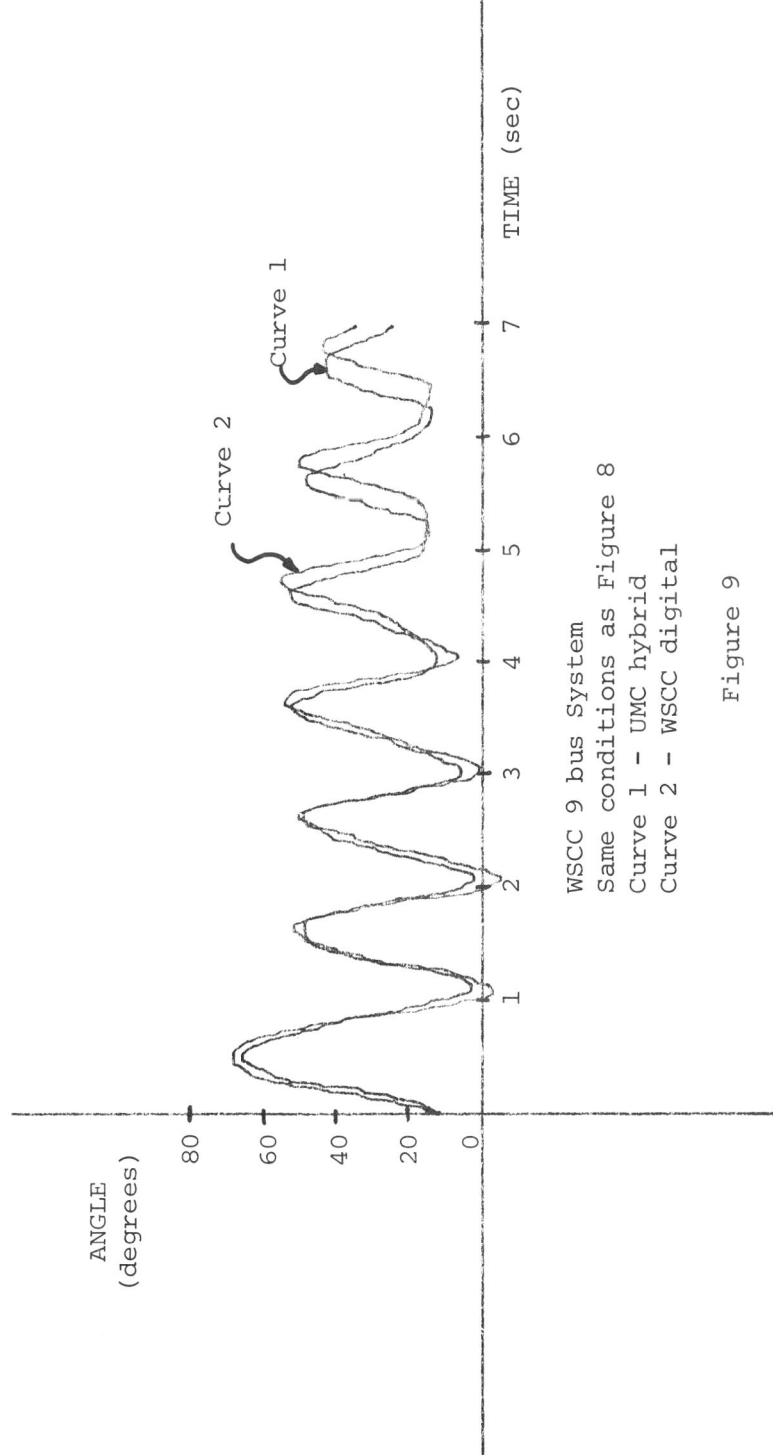

Figure 9

2. Long Term Dynamics (9)

A test system consisting of 16 generation units and 68 busses was selected for preliminary long term dynamics tests on the hybrid system. This system was provided by General Electric and was utilized by both RP907-1 and RP908-1. Several long term dynamic scenarios were run on the 16 machine system. One of the scenarios run caused the 16 machine system to break into four islands.

The frequencies for two of the islands started to go opposite directions. Under this condition LRM position (MW demand) control for the generation units in these two islands are ramped, in a closed-loop fashion, in an attempt to bring their corresponding island frequencies back to normal. As a consequence the two islands were resynchronized at 13.5 minutes after the break up.

Figure 10 shows the resulting frequencies of the two islands from t=0 to t=1200 sec. (20 minutes). This scenario demonstrates the resynchronizing capability of the hybrid simulator. The events and data collected during this scenario are fully described in the RP908-1 final report.

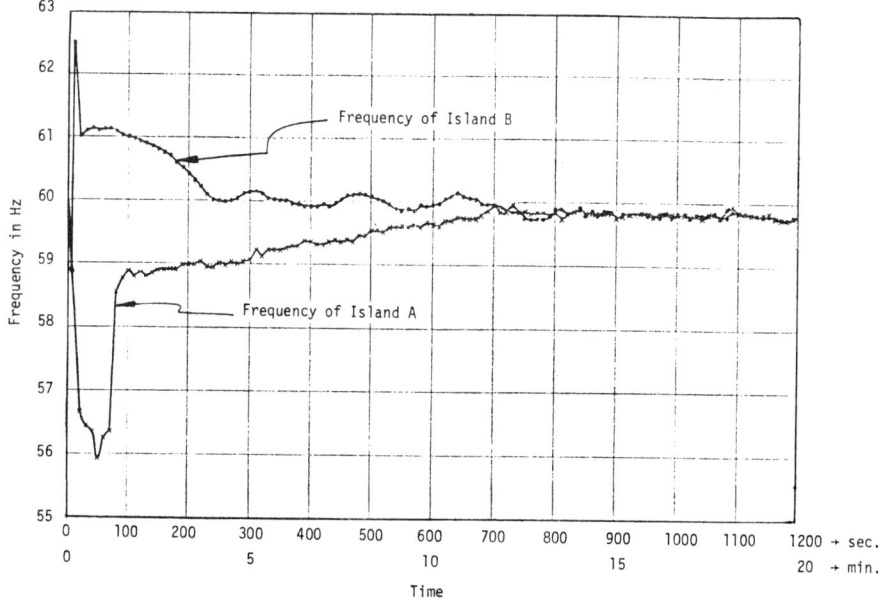

FIGURE 10
Island Frequencies vs Time (0 - 20 min.)

VII. HYBRID SIMULATOR RELAY IMPLEMENTATION (4, 5)

The hybrid simulator has a relaying package consisting of underfrequency relays, overcurrent relays, and impedance relays. The underfrequency and overcurrent relays are primarily implemented in hardware, whereas the impedance relays are implemented in software.

The one line diagram of the one machine infinite bus system shown in Figure 11 is utilized for relay tests. The tests run on this system are grouped in the following two categories by the system condition that was varied:
1. Generator terminal voltage level
2. Fault duration.

With these system operating conditions as variables, the system reaction to a fault at Bus 28 was observed. After a predetermined time, the fault was removed by dropping line 28-29. For each test, the impedance as "seen" by a directional impedance (MHO) relay at Bus 29 on line 29-26 was observed. The fault at Bus 28 does not fall within this relay operating characteristic and only system swings resulting from the fault could cause this relay to operate.

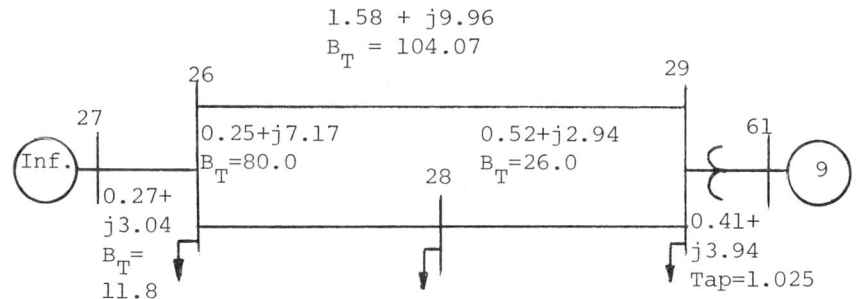

P_{g9}=408.0 MW Q_{g9}=60 MVAR
V_{27}=1.000 p.u.
Load 26   P = 288.7 MW   Q = 27.0 MVAR
Load 28   P = 285.5 MW   Q = 34.8 MVAR
Load 29   P = 264.3 MW   Q = 72.0 MVAR

One-Line Diagram of Test System

Figure 11

The R-X Diagram of $\overline{Z}_{D-B}$ (Figure 12) illustrates the variation in impedance as a function of time beginning after the application of the fault. The relay operating characteristic is of a MHO relay set for 100% of the line length at the line angle. The impedance as seen without loss-of-synchronism ($|\overline{V}_{61}| = 1.06$ p.u.) does not pass through the operating characteristic of the relay. As the prefault voltage $|\overline{V}_{61}|$ decreases, the minimum value of $|\overline{Z}_{26-29}|$ decreases and the speed of the slip cycle increases, as illustrated by the fewer points on the curve for $|\overline{V}_{61}| = 0.99$ p.u. compared to $|\overline{V}_{61}| = 1.04$ p.u. The data points are approximately 0.075 seconds apart. The discontinuities (dotted lines for these curves are the point at which lines 28-29 opened removing the fault.

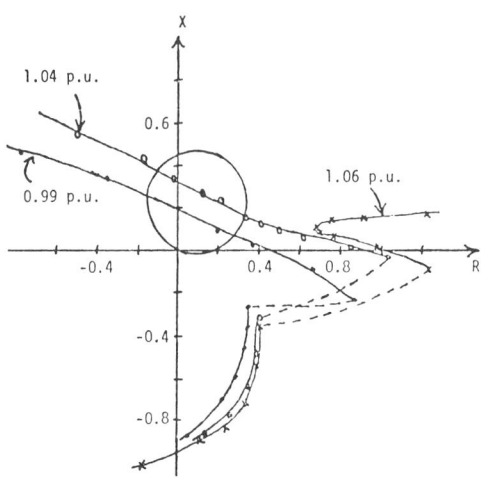

One Machine - Infinite Bus Test System -
Impedance for 3φ Fault @ Bus 28 - Generator
Voltage ($|\overline{V}_{61}|$)

Figure 12

The underfrequency relaying hardware monitors an analog signal to detect whether or not the frequency has fallen below the UNDERFREQUENCY SET POINT. A counter measures the length of time during which the signal is under the frequency set point. If this length of time exceeds the value set in the TIME DELAY LIMIT register, than an UNDERFREQUENCY SIGNAL is triggered. The UNDERFREQUENCY SET POINT and TIME DELAY LIMIT registers are loaded from the computer.

The overcurrent relay implemented in hardware on the long term dynamics hybrid simulator monitors the transmission line current. The current input is a dc voltage produced by an ac measurement circuit and filtered through a variable time constant filter and compared with a reference value. If the value of current exceeds the reference value for a certain period of time (determined by the variable time constant filter), then the relay operates.

The variable time constant filter ranges between approximately 200 ms and .5 seconds (or in real time, 4 ms to 10 seconds).

The digital computer scans the breaker signals from each overcurrent relay. When any of the breaker signals are high, then the computer opens the analog switches to remove the appropriate transmission line from the power system.

The impedance relay is primarily implemented in software. The inputs to the program from the analog system are voltage magnitude, current magnitude, and real and reactive power from the protected line. The impedance data is calculated by the program as follows:

$$\text{Impedance magnitude} = \frac{\text{voltage magnitude}}{\text{current magnitude}}$$

$$\text{Impedance angle} = \tan^{-1}\left(\frac{\text{reactive power}}{\text{real power}}\right)$$

The minimum sample period for impedance relaying is about 5 ms of 1200 Hz time or 0.1 second (6 cycles) on 60 Hz time. The impedance can be plotted on a CRT or an XY plotter with data points spaced in time at 0.1 second intervals (60 Hz time).

VIII. CONCLUSIONS

The recent developmental projects at the University of Missouri - Columbia has shown that it is possible to develop low cost hybrid simulators to be used in the electric power industry.

The present operational capability of the faster-than-real-time, long term dynamics hybrid simulator can be summarized as follows:
1. Running long-term system dynamics studies
2. Performing system stability studies
3. Running load flow studies

4. Performing sensitivity studies of various parameters
   a. generator, exciter and regulator coefficients
   b. load coefficients
   c. governor characteristics
   d. effects of transformer saturation
5. Providing flexibility to determine corrective measures, control parameters and operating procedures.

It can be concluded that:
- The hybrid approach is a viable simulation technique.
- Can have fast start up.
- Auto Patching is feasible.
- Computer controlled parameter settings are feasible.
- The hybrid simulator results compare closely with other accepted simulation tools.

Highlights of accomplishments of the RP908-2 project can be summarized as follows:
1. A full two axis generator model works very well using hybrid techniques with the AC model.
2. The intermediate time results (10-100 sec) are very sensitive to the load characteristics on both analog and digital models.
3. The system with 4000 amplifiers has been on continuously 24 hours per day with an average of 10-12 hours per day operation for 2453 hours with 7 failures, 5 analog and 2 digital computer or peripheral. These failures caused the system to be down for 8.75 hours for a system availability of 99.64%.
4. Rapid synchronization was demonstrated.
5. It was shown that it is feasible to have all major parameters under computer control.
6. Computer controlled patching is economically feasible for this type system.

It is concluded that the hybrid simulator can be used as a research tool for the study of dynamics and different modeling approaches. Since the system solution time is independent of system size, many more studies can be performed in a shorter time than is practical with the existing digital programs. This allows a much wider range of sensitivity tests to be made than has previously been practical. This will allow the various control systems to be further studied and possibly redesigned to enhance system operation.

A hybrid simulator can be used in-house by utilities for

a variety of planning and contingency studies. Detailed sensitivity analysis on large power systems can be performed which are now virtually impossible using the large digital programs. The sensitivity analysis is important to further determine such things as coordinated control design or exciter design and settings. This information may be valuable to determine operating procedures under certain conditions when sensitivity of certain parameters is greater. Another area of great potential for the hybrid simulator is an on-line security monitor. The hybrid approach is required to obtain the results of contingencies in real time or faster-than-real-time solution which is necessary before closed loop control can be performed. In addition it is felt that the simulator will prove valuable as an aid to develop dynamic equivalents and for dispatch operator training for large interconnected systems.

IX ACKNOWLEDGEMENTS

This research was sponsored primarily by the Electric Power Research Institute under project RP908-1 and RP908-2, with Timothy S. Yau as the project manager.

X. REFERENCES

(1) Lyons, A. J., "High Speed, AC Simulation of Electrical Power Systems", Ph.D. Dissertation, University of Missouri - Columbia, 1971, pp. 2-4.
(2) Warshawsky, M. L., "Background for the Advanced Hybrid Computer System", Special Symposium on Advanced Hybrid Computing Proceedings, pp. 149-156, July 1975.
(3) Saucier, A., "An Overview of Advanced Hybrid Computing", Special Symposium on Advanced Hybrid Computing Proceedings, pp. 157-160, July 1975.
(4) Wolin, L., "The Wolin-Saucier-Peak (WSP) Scientific Mix, A Quantitative Method for Comparing Hybrid and Digital Computer Performance", Special Symposium on Advanced Hybrid Computing Proceedings, pp. 181-188, July 1975.
(5) Hutchinson, C. E., and Seshachar, N., "A Note on the Cost of Digital Simulation", Special Symposium on Advanced Hybrid Computing Proceedings, pp. 189-190, July 1975.
(6) Sorondo, V. J., Wavell, R. B., and Nixon, F. E., "Cost and Speed Comparison of Hybrid vs Digital Computers", Special Symposium on Hybrid Computing Proceedings, pp. 191-197, July 1975.

(7) Peak, D. C., "A Survey of Requirements for Present-Day and Fourth-Generation Hybrid Computer", Special Symposium on Advanced Hybrid Computing Proceedings, pp. 161-166, July 1975.

(8) Michaels, L., "Investigations of Hybrid Systems for Power Systems Analysis, The AC/Hybrid Power System Simulator", EEI RP90-1 Final Report, prepared by Electronics Associates, Inc. for the Edison Electric Institute, November 1970.

(9) University of Missouri - Columbia, "Long Term System Dynamics - Hybrid Simulation", EPRI RP908-1 Final Report, prepared for the Electric Power Research Institute, May 1975.

(10) Carroll, D.P., "Digital and Hybrid Computer Simulation of Power Systems", Simulation, July 1973, pp. 9-15.

(11) Krause, P.C., "Application of Analog and Hybrid Computers in Electric Power Research", Simulation, August 1970, pp. 73-79.

(12) Carlson, N.R. and Hoffman, A.G., "Power System Security Evaluated by Hybrid Load Flow Calculator Using On-Line Data", Midwest Power Symposium Conference Record, 1970.

(13) Enns, M., Giras, T.C. and Carlson, N.R., "Load Flows by Hybrid Computation for Power System Operations", 1971 PICA Proc., pp. 401.

(14) Michaels, L., op. cit.

(15) Michaels, L., "The AC/Hybrid Power System Simulator and Its Role in System Security", 1971 PICA Proc., pp. 408.

(16) University of Missouri - Columbia, "Design and Operation of the Consolidated Edison Analog Over Voltage Simulator (AOVS)", A Final Report prepared for the Consolidated Edison Company, 1975.

(17) Walker, L.N., Jennings, J.B., Ott, G.E., and Ho, L.Y., "Design and Operation of an Electronic Analog Overvoltage Simulator", presented at IEEE PES Winter Meeting, January 1975.

(18) Ott, G.E., Walker, L.N., and Wong, D.T.Y., "Hybrid Simulation for Long Term Dynamics", IEEE Transactions on Power Apparatus and Systems, Vol. PAS 96, May/June 1977, pp. 907-915.

(19) University of Missouri - Columbia, "Hybrid Simulation of Power Systems Dynamics - Feasibility Study", EPRI RP908-2 Final Report, prepared for the Electric Power Research Institute, February 1976.

(20) Anguvorakul, Somchai, "Hybrid Simulation of Power Generating and Transformer Units for Long Term Dynamics studies", PH.D. Dissertation, University of Missouri - Columbia, May 1975.
(21) Markovic, Miroslav, "Mathematical Derivation of Synchronous Generator Parameters", Technical Note, EPRI RP908-2, University of Missouri - Columbia, December, 1975
(22) Wong, Daniel Tai-Yui, "Computerized Patching for Power System Hybrid Simulator", Ph.D. Dissertation, University of Missouri - Columbia, May 1976.

Power Control System and Protection

ELEMENTS OF POWER SYSTEM IDENTIFICATION AND CONTROL

Bruce K. Colburn

Texas A&M University

I. INTRODUCTION

The field of estimation and identification can be traced back to the work of Karl Friedrich Gauss, who at the age of 18, developed the least squares method in 1795 for planet and comet motion [1]. Much has happened during the ensuing 180 years, including the development of the concepts of maximum likelihood estimation, Wiener filtering, stochastic approximation, Kalman filtering, and model-reference adaptive identification to name just a few of the techniques developed over the years. In the area of parameter identification, the demands of WW II fighter aircraft triggered the modern interest by control engineers in system parametric identification. During the 1950's-60's applications areas expanded in the industrial and aerospace sectors. This period also witnessed a surge in interest by systems engineers and a great deal of theoretical development occurred. From the late 60's on, the emphasis has been on the practical development and implementation of such algorithms. Some of the areas now witnessing the application of such methods include the chemical industry, human operator tracking, econometric dynamic models, multi-unit manufacturing production lines, biological, high performance aerospace areas, as well as power system studies.

In the past, the power industry has historically relied upon well-proven, conservative design and analysis techniques for determination of load projections, equipment requirements, stability and control techniques, and overall system operation. However, as operating costs and expansion needs rise at a rate faster than those of utility revenues [2], it becomes important to look at new and more innovative ways of obtaining greater use of existing and new facilities than has been done in the past, and still insure safe operation. Identification of significant system parameters which were unknown or not well-defined is one of these ways of improving system performance at minimum cost. In fact, during the last few years a great amount of work has been done on applying the techniques of modern control theory to the operation, modeling, analysis, and synthesis of the various phases of power operation. Examples include (a) load flow studies, (b) system static and dynamic equivalencing [3], (c) tie line power and frequency control [4], (d) system state estimation [5],

(e) automatic generation control [6], (f) unit commitment [7], (g) system stability [8], (h) transient machine control [9], (i) optimal control of power systems [10], (j) load-frequency control [12, 13], (k) machine modeling and identification [14], (l) transient stability studies [15], generator dynamics identification [16], and power dispatch [17].

Because of large-scale network interconnections, problems of area power control (APC) and area control error (ACE) are magnified, load-frequency control is complicated, and interconnection grid stability is more critical. It becomes more important than ever to obtain system information as is available from dynamic equivalents, in advance, concerning potential system responses due to a fault or disturbance in a portion of an interconnection. Generator operating curves vary with load and for efficient power control, knowledge of the time-varying static and dynamic operating characteristics is required. Using identification techniques such information can be acquired.

Both parametric and non-parametric identification is possible, but virtually all modern systems work, and the needs of the power industry, fall under the first heading. Parametric identification means that one is interested in learning the value of significant power equation coefficients, so as to obtain analytical models which can then be used for signal estimation, security assessment, control, etc. Nonparametric methods lead primarily to impulse response, Fourier analysis, and correlation techinques. Offline methods are applicable when sufficient time is available to gather a large quantity of measurement data, and then manipulate the data (the identification algorithm) so as to obtain parameter estimates. Examples would include least squares, maximum likelihood, and crosscorrelation. One of the drawbacks of these approaches is that a basic assumption of system parameter time-invariance over the observation period is required, an assumption which does not hold for many power system combinations. Online identification techniques operate in real-time (except for a possibly brief updating time), providing a continuously updated set of parameter estimates. Two basic differences between these approaches, then, are (1) off-line methods provide a single model with constant parameters, whereas the latter exhibits a continuously revised model with possibly (slowly) time-varying coefficients, and (2) off-line data is available only after some large time period T (minutes to hours) has elapsed. These principles are illustrated in Figure 1.

Even though most power system component dynamic characteristics are nonlinear in nature, resulting experience to

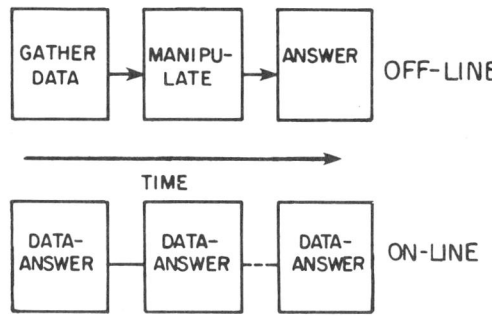

Fig. 1. Concept of On-Line versus Off-Line Identification

date suggests that it suffices in most cases to obtain linearized equivalent models. This is fortunate since most identification algorithms are designed for linear, time-invariant situations. The purpose of this chapter is to overview the application of systems identification to the modeling and control of linear dynamical power systems problems of the present and future. A few illustrative numerical results are included to illustrate the application of identification theory to power system operation.

II. THEORY OF POWER SYSTEM IDENTIFICATION

From the different components making up a power system (motors, generators, transmission lines, etc.), a series of math models describing both the static and dynamic characteristics can be developed. Static equation models might be of the form [18]

$$P_{ik} = E_i E_k Y_{ik} (\delta_i - \delta_k) \tag{1}$$

which represents the inter-power flow between two network busses, for a boundary flow bus network model, where P_{ik} is real power flowing from bus i to k, δ_i is a voltage phase angle at bus i, E_i is the voltage magnitude at bus i, and Y_{ik} is the transfer admittance between busses i and k. In (1), Y_{ik} would generally be regarded as a system parameter of interest. Other examples would include

$$\underline{I} = Y \underline{E} + \underline{K} \tag{2}$$

a set of algebraic equations specifying phasor relationships between current and voltage at a selected set of generators and terminal nodes of a network, with $\underline{I}$ the terminal and generator node currents, $\underline{E}$ the terminal and generator node voltages, Y the constant admittance matrix, and $\underline{K}$ a constant vector. Y and $\underline{K}$ would be the parameter terms of interest.

Examples of other power system parameters of interest include transmission line susceptance and conductance, and transformer tap ratios and admittances.

A dynamical model of a system would be of the general form

$$\underline{\dot{x}} = \underline{f}(\underline{x}, \underline{u}, t, \underline{\omega}) \qquad (3)$$

where $\underline{x}$ are the system states, $\underline{u}$ the set of deterministic exciting inputs, t is the running time variable, and $\underline{\omega}$ is a vector of uncontrollable, unmeasurable inputs representing system noise or disturbances. Examples of system states would include phase angles, generator output voltages, power flow, and line frequency. Examples of $\underline{u}$ would include turbine steam pressure, firing control signals, prime mover torque increments, or exciter voltage error. By linearizing (2), the general equation

$$\underline{\dot{x}} = A(t)\underline{x} + B(t)\underline{u} + \Gamma(t)\underline{\omega} \qquad (4)$$

results, where $A(t)$, $B(t)$, $\Gamma(t)$ are time varying in general. It is equations of the form of (3) that are of interest here.

As an example of dynamic power system modeling, consider the system shown in block diagram form in Figure 2. Significant parameters of interest include the effective rotary

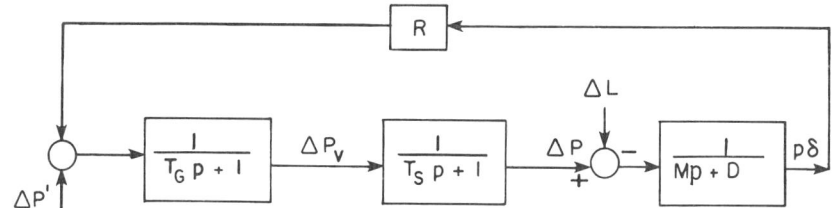

Fig. 2. Model of a Power-Speed Governor For A Generator

inertia (M), damping torque coefficient (D), governor time constant (T_G), and turbine time constant (T_S). The controllable inputs may include $\Delta P'$, the change in valve position (speed changer position), ΔL the area load change, ΔP the incremental prime-mover torque produced by governor-controlled valve motion ΔP_V, and R is a speed regulation term.

As given by Zadeh [19], an identification problem is characterized by three quantities, 1) a class of models, 2) class of input signals, 3) a criterion. Basically, the process evolves in three stages,

1) hypothesize a reasonable model
2) estimate parameters of that model using some numerical algorithm operating on the measurable input-output data
3) test the validity of the resulting model

Most of the effort in this field has been towards algorithm development, with only limited work on modeling [20, 21]. The determination of algorithms for identification has covered the fields of statistics, mathematics, and engineering, as all have had a need for parameter determination.

In practice, the models used for identification will be of lower order than the actual system since the system is of infinite order. The key problem then becomes one of determining the best parametric model whose parameters have a significant relationship to the actual system parameters. Testing of "goodness-of-fit" is usually effected by determination of mean-square error, statistical measures, and individual evaluation of the facts. Some of the performance indices (IP) used include 1) minimization of a norm of generalized output error, $||\varepsilon||$, 2) minimization of the state errors, $||\underline{x}_m - \underline{x}_p||$, 3) and minimization of the parameter ($\underline{\theta}$) structure distance, $||\underline{\theta} - \underline{\hat{\theta}}||$. It is the third case of theoretical concern, but it is also the one IP that is not physically measurable, for if $\underline{\theta}$ were known, why identify it? This identification evaluation principle is illustrated functionally in Figure 3.

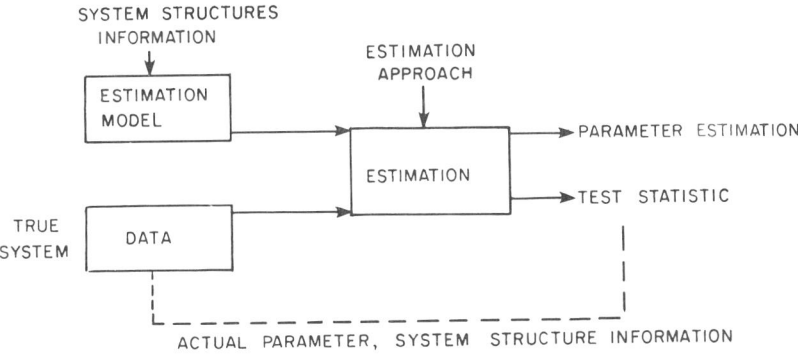

Fig. 3. Identification - Modeling Problem

Because "real-world" power systems are of infinitely high order, it is often necessary in practice to determine a "best" low order (i.e. 2 or 3) model which, according to some cost function, best describes the real (power) system. Of course, by increasing the model order the fit will be improved. To test whether the improvement is statistically significant one way is to use the loss function [22]

$$V = \sum_{k=1}^{n} e(k)^2 \qquad (5)$$

and test

$$t = \frac{V_1 - V_2}{V_2} \cdot \frac{N - n_2}{n_2 - n_1} \qquad (6)$$

where n_1 = number of parameters for case 1, n_2 = number of parameters for case 2; V_1 and V_2 are the minimum values of the loss function for n_1 and n_2, and N is the number of data points. It can be shown that at the 5% risk level, $t > 3$ for significance, which provides a decision criterion for model selection.

It should be noted that in dynamic system identification there are two main types of identification, 1) state identification and 2) parameter identification. The state, or output, identification case for a given input $\underline{u}(t)$ does not assure identification for all possible plant inputs, and is most suitable for the model-matching control situation. Parameter identification is of most interest here.

When the plant input is "sufficiently exciting," transfer function (or transfer matrix) identification can be assured. However, there are situations that exist for which transfer function identification is not equivalent to plant parameter identification. To somewhat alleviate this problem, identification methods are generally predicated on a particular plant parameter structure, and only parameters in the structure are identified. Sometimes additional analytical relationships are then required to relate identification parameters to the physical plant parameters. Fortunately, in most power system cases, such low order models are utilized for identification that direct physical plant parameter identification is feasible.

III. DYNAMIC EQUIVALENCING

Of special interest in power systems are means of determining simplified or decoupled models for the actual large-scale interconnected power systems. Much of the work in large-scale systems addresses the problem of partitioning to obtain localized equivalents with coupled external components. Dynamic equivalents are a compromise approach, where a local system is represented in detail, but the coupling terms represented by the remaining external system are treated as a low order model excited by noise (representing non-controllable external inputs). This "equivalencing" then constructs

a compact and adequate mathematical representation for the
external system that accurately reproduces its impact upon
the local area.

Equivalents have small computer storage and time requirements. They can be used in different system studies, where
the area of concern is represented in detail and the external
system expresses its response for disturbances originated in
the locally controlled area through the reduced order equivalents. An electric power system interconnection is divided
into local and external systems. The boundary busses are
defined as shown in Figure 4. The choice of the boundary
busses is such that any large disturbances to be investigated

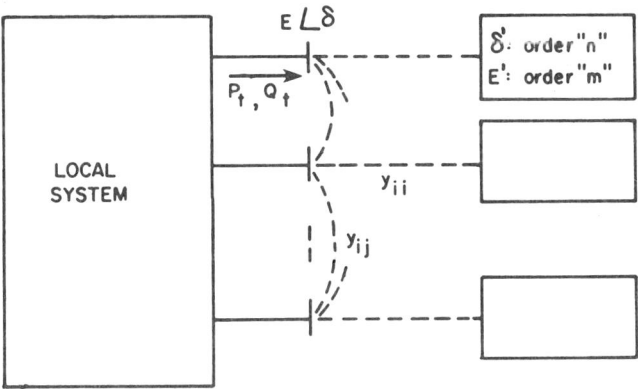

Fig. 4. The External System-Dynamical Equivalent Model Concept

in the local system would have small effects on the external
system. This allows the assumption that the external system
can be represented by a time invariant linear model during a
simulation study. During this period, the external system
can be viewed under balance at an operating point where the
sum of the flows into the boundary busses makes for the
difference between total load and total generation.

To represent the quasi steady-state of the external
system, it is replaced by its equivalent admittance matrix
as seen from the boundary busses. This could be constructed
by formulating the bus admittance matrix, then reducing it
by eliminating busses retaining the boundary ones. For constructing the external system dynamic equivalent, the external system is replaced by a number of fictitious busses
extended through the tie lines across the boundary busses of
the local system. Each of these busses is viewed as independent, and collectively they convey the impact of the external
system upon the local one in response to disturbances within
the local system. The procedure consists of [3]

a) collecting data on the tie lines connecting local and external systems

b) construction of the matrix of driving point and transfer admittances of the external system as seen from the boundary busses

c) computation of fictitious bus values corresponding to the measured ones on the boundary busses

d) identifying key system parameters in the appropriate dynamic model equations.

Under the assumption of large disturbances to be locally limited with minimal impact on the external system, then the external system can be viewed as a stochastic process excited by unknown disturbances and observable inputs. An accepted model form is the ARMA model

$$Y(k) = \sum_{i=1}^{n} \alpha_i Y(k-1) + \sum_{i=1}^{m} \alpha_{n+i} W(k-i)$$

$$+ \sum_{j=1}^{r} \sum_{i=1}^{h} \gamma_{ji} U_j (k-i) + W(k) \qquad (7)$$

where $Y(k)$ represents the changes in voltage magnitude and phase angle, $W(k)$ the residual error of $\underline{Y}$, $U_j(k)$ are the changes in tie lines active and reactive power n,m,h the model order, r number of input variables affecting the value of Y, and α_i, β_i model parameters.

Estimating the dynamic equivalent therefore requires a sequence of measurements of voltages, and phase angles, and tie line active and reactive power at each of the boundary busses. The external system representation at each of the boundary busses is shown in Figure 5, where $P_t(k)$, $Q_t(k)$ represent the inflow of power to boundary busses from local to external system; $P_i(k)$, $Q_i(k)$ represent the outflow of power from the ith boundary bus through assumed admittances to the ith fictitious bus; Y_{ij}, Q_{ij} represent the ij tranfer admittance of the external system as seen from the boundary busses of the local system; $P'_i(k)$, $Q'_i(k)$ represent the inflow of power to fictitious bus i at instant k; $E_i(k)$, $\delta'_i(k)$ are the voltage magnitude and phase angle on fictitious bus i at instant k; $\bar{E}'_i$, $\bar{\delta}'_i$ are the steady state values of voltage magnitude and phase angle on fictititous bus i; $\Delta E'_i(k)$, $\Delta \delta'_i(k)$ are deviations of voltage and phase angle from the steady state value on fictitious bus i.

For each equivalent internal source, two models are identified, one for voltage and one for phase angle, i.e.

$$\Delta E_i'(k) = \sum_{j=1}^{n} \{\alpha_j \Delta E_i'(k-j) + \beta_{1j} \Delta P_i'(k-j) + \beta_{2j} \Delta Q_i'(k-j)\} + \sum_{j=1}^{m} \alpha_{n+j} W_E(k-j) \quad (8)$$

and

$$\Delta \delta_i'(k) = \sum_{j=1}^{n} \{\alpha \Delta \delta_i'(k-j) + \beta_{1j} \Delta Q'(k-j)\} + \sum_{j=1}^{m} \alpha_{n+j} W_\delta(k-j) \quad (9)$$

The parameters α_i and β_i may then be identified for different models by varying $(n,m,\tilde{n})$. Using statistical tests, the "best" low order fit models may then be determined.

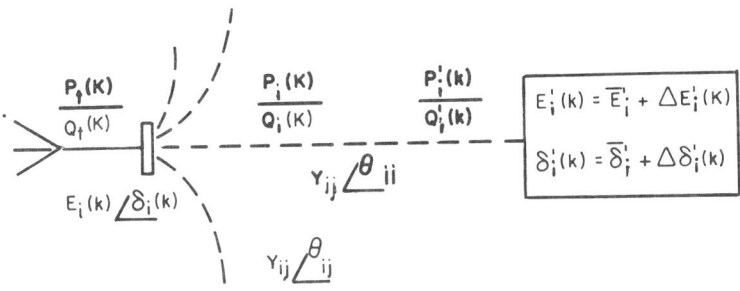

Fig. 5. Identifying Equivalents at Each Boundary Bus

IV. IDENTIFICATION APPROACHES

There are a number of identification approaches which have been developed, some of which have been applied to power systems. They include generalized least-squares [23], least squares, maximum likelihood [24], Bayesian, model reference adaptive control (MRAS) change [25-28], instrumental variable [29], cross correlation [30], extended Kalman filtering [31], and frequency-domain [32]. The frequency-domain approach is most familiar to engineers, as the frequency response data is used to define gains, poles, and zeroes of the input-output response, represented by Bode plots. In practice, however, such an approach is not practical for power systems due to lack of input control freedom. The concept of parameter identification from input-output measurements follows along the same principles, however.

To illustrate the input-output identification concept, consider the discretization of (4), which can then be written in recursive form as

$$Y(j) + \sum_{i=1}^{n} a_i Y(j-1) = \sum_{k=1}^{r} b_k u(j-k)$$
$$+ \sum_{\ell=1}^{m} \gamma_\ell \omega(j-\ell) + \omega(j), \quad j=0,1,2,\ldots \quad (10)$$

Neglecting the noise terms, this could represent the transfer function

$$T(z) = \frac{b_1 z^{-1} + b_2 z^{-2} + \ldots + b_r z^{-r}}{1 + a_1 z^{-1} + a_2 z^{-2} + \ldots + a_n z^{-n}} \quad (11)$$

if a_i and b_i are assumed time invariant. The a_i and b_i are the system parameters which determine the transient time response of the system. Of course, in practice, failure to consider the noise will generally result in biased estimates.

One of the most widely used approaches is the least-squares parameter estimation of the autoregressive, moving average (ARMA) system (4) [33] where one performs the operation

$$\text{Select } M_\ell: \quad J = \min \left\{ \sum_{j=1}^{N} [\underline{y}(j) - \hat{\underline{y}}(j,\underline{\theta})]^T \Lambda^{-1} \right.$$
$$\left. [\underline{y}(j) - \hat{\underline{y}}(j,\underline{\theta})] \right\} \quad (12)$$

for some hypothesized model M_ℓ, where $\hat{\underline{y}}$ is the estimated output, Λ is a weighting matrix, and N is the number of discrete data points used. The purpose of introducing a selection procedure (M_ℓ) into the criterion is to allow for a comparison of models which differ in structure, i.e. order of system, transportation lag, etc. Numerous examples for SISO systems are given in [34].

Using (10) to represent the model of linear process (with $\omega(\ell) \triangleq 0$, $\ell=0,1,2,\ldots$), the estimation error

$$e(j) = y(j) - \hat{y}(j) \quad (13)$$

must be minimized. For the classical least squares case, Λ^{-1}

in (12) is I_n, and J represents the loss function

$$V = \sum_{j=1}^{N} e^2(j) \tag{5}$$

To determine $\hat{\underline{\theta}} = [\hat{a}_1 \ \hat{a}_2 \ldots \hat{a}_n \ \hat{b}_1 \ \hat{b}_2 \ldots \hat{b}_r]^T$, the parameter estimate is

$$\hat{\underline{\theta}} = [\psi^T \psi]^{-1} \psi^T \underline{Z} \tag{14}$$

where $\underline{Z} = [y(r) \ y(r+1) \ldots y(r+N)]^T$ and

$$\psi = \begin{bmatrix} -y(n-1) & -y(n-2) & \ldots & -y(o) & | & u(r-1) & u(r-2) \ldots u(0) \\ -y(n) & -y(n-1) & \ldots & -y(1) & | & u(r) & u(r-1) \ldots u(1) \\ & & & & \vdots & & \\ -y(n+N-1) & -y(n+N-2) & \ldots & -y(N) & | & u(r+N-1) & \bullet \bullet \bullet \quad u(N) \end{bmatrix} \tag{15}$$

Variations on the least squares techniques for so-called "on-line" techniques have also been developed [35], although they are not truly on-line. A finite time-span for data collection as in (15) is required, then a series of "frozen-time" estimates of $\underline{\theta}$ are obtained. The recursive least squares can be written as [36]

$$\hat{\underline{\theta}}(k+1) = \underline{\theta}(k) + [\psi(k+1) \ P(k) \ \psi^T(k+1) + 1]^{-1}$$
$$P(k) \ \psi^T(k+1) \ [\underline{y}(k+1) - \psi(k+1) \ \hat{\underline{\theta}}(k)] \tag{16}$$

$$P(k+1 = P(k) \ [-\psi^T(k+1) \ \psi(k+1) \ P(k)$$
$$[\psi(k+1) \ P(k) \ \psi^T(k+1) + 1]^{-1}] \tag{17}$$

with

$$P = [\psi^T \psi]^{-1}$$

Other methods depend on minimization of a functional such as

$$V = \int_o^T e(t)^2 dt \tag{18}$$

where T is large enough to include the significant effects from the transient identification. Most are iterative, such as [22]

1. stochastic approximation

$$\underline{\theta}(k+1) = \underline{\theta}(k) + \alpha(k) \ V_\theta[\underline{\theta}(k)] \tag{19}$$

with $\alpha(k) = \frac{1}{k}$ as an example

2. gradient method

$$\underline{\theta}(k+1) = \underline{\theta}(k) + \alpha \, V_\theta[\underline{\theta}(k)] \qquad (20)$$

$\alpha > 0$, constant

3. steepest descent

$$\underline{\theta}(k+1) = \underline{\theta}(k) - \alpha(k) \, V_\theta[\underline{\theta}(k)] \qquad (21)$$

where $\alpha(k)$ is optimized with respect to the gradient direction. The advantage of the on-line techniques is that some estimate of $\underline{\theta}$ is available in a short time (however inaccurate it may be). For many power-control problems a best estimate is better than none. Some comparative studies of various identification approaches have been done [36-38], but exhaustive results are few. For some surveys of the aforementioned and other like techniques, see [22, 39, 40].

An on-line identification technique which insures parameter convergence as part of the algorithm is that of the model-reference adaptive systems (MRAS) [25-28] approach. This method is a significant departure from traditional approaches in a number of ways, (1) a scalar metric $\lim_{t \to \infty} e(t) = 0$ or $\lim_{t \to \infty} E\{e\} = 0$ is used as opposed to scalar or vector summation or integral criteria, (2) using Lyapunov theory it is possible to insure global stability of the parameter identification, (3) the MRAS technique is inherently an on-line sequential approach, allowing for its use in the identification of nonstationary systems, (4) unknown system parameters are treated as "gains" of a model, with the end result being the matching of model parameters to those of the system.

The basic MRAS approach is illustrated in Figure 6. There is need for input-output measurement data from a system, which is then manipulated mathematically in a computer, the result of which are the coefficients of a model, which in some sense represents the physical problem. Computational requirements are generally minimal for low-order systems and the MRAS approach is easily amenable to minicomputer use, such as may be found in existing substation hardware. The utility of this approach for power systems problems are (1) MRAS can handle time-varying systems, (2) approximate results are available in real-time, where data can be used for control, and (3) model coefficients are generated directly.

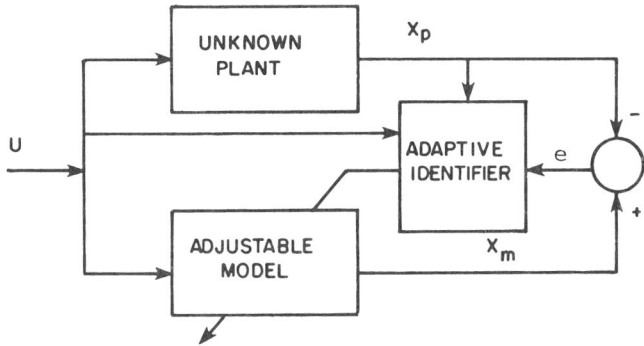

Fig. 6. Model-Reference Adaptive System (MRAS) Identification Principle

A number of different MRAS identification structures have arisen as a result of MRAS study, including the parallel, series-parallel and series methods. These refer to the ways in which adaptation of a variable model is effected from input-output plant information. The parallel case corresponds to the "output error" method in recursive identification and series-parallel to the "equation error" approach. Three key means by which identification can be effected include 1) direct parameter adaptation, 2) signal synthesis adaptation, whereby a controlled auxiliary input is applied to the system, and 3) combination of parameter and signal-synthesis adaptation. The one of most interest in power identification is #1, and it is this one to which most of the work has been addressed.

Various MRAS algorithms are available, one of which for the (noise-free) system of Eq. (4) is [28]

$$a^m_{n_j} = \alpha_j \int_{t_o}^{t} v_{f_1} x_{m_{f_j}} \, dt - \beta_j (v_{f_1} x_{m_{f_j}}) \qquad (22)$$

$$b^m_j = -\alpha_j \int_{t_o}^{t} v_{f_1} u_{f_j} \, dt - \delta_j (v_{f_1} u_{f_j}) \qquad (23)$$

where $a^m_{n_j}$, b^m_j are the parameter estimates for A, B in phase-variable form, v_{f_1} is a filtered error between the plant and model, x_{mf_j}, u_{f_j} are filtered model data, and α_j, $\delta_j > 0$, and β_j, $\delta_j > 0$.

An important problem to be solved for MRAS on-line use is the area of convergence rate. This is especially important for time-varying on-line parameter identification. Some success to date has been achieved using error equations [27] and Floquet Theory [41, 42]. More work remains to be done, however, before one can accurately predict parameter convergence rates for any general adaptive observer configuration.

Another method which has met with widespread acceptance is the maximum likelihood (ML) technique, based on maximizing (in a probabilistic sense) a function involving the unknown parameter vector $\underline{\theta}$. Given the system

$$y(k) = \sum_{i=0}^{n-1} a_i \, y(k-n+i) + \sum_{i=0}^{n-1} b_i \, u(k-n+i)$$

$$+ \sum_{i=0}^{n} c_i \, N(k-n+i) \quad (24)$$

where $\sum_{i=0}^{n} c_i \, N(k-n+i) = \sum_{i=0}^{n} [d_i w(k-n+i) - a_i \, v(k-n+i)$ and $w(k)$, $v(k)$ are zero mean with variances W^2, V^2, it is desired to identify the 3n-parameter vector $\underline{\theta}^T = [a_1 \, a_2 \ldots a_n, \, b_1, \, b_2 \ldots b_n, \, c_1, \, c_2 \ldots c_n]$. The output error is

$$e(k/\underline{\theta}) = z(k) - \hat{z}(k/\underline{\theta}) \quad (25)$$

with variance $\sigma_e^2 = E\{e^2(k/\underline{\theta})\}$. The idea is to maximize $L_K(\underline{\theta}, \sigma_e)$ WRT $\underline{\theta}, \hat{\sigma}_e$, where

$$L_K(\underline{\theta}, \sigma_e) = \ell n \, [p(z(k)/u(k-1); \underline{\theta})] \quad (26)$$

It can be shown [36] that an on-line version of ML is as follows:

$$\hat{\underline{\theta}}(N+iM) = \hat{\underline{\theta}}(N+(i-1)M) - \frac{1}{\big((N+iM)/(M+1)\big)} S^{-1} R \quad i=1,2,\ldots \quad (27)$$

where M is the discrete number of updating periods used,

$$S = \sum_{j=N+1}^{N+M} \left[\frac{\partial e(j/\underline{\theta})}{\partial \underline{\theta}} \cdot \frac{\partial e^T(j/\underline{\theta})}{\partial \underline{\theta}} \right] \quad (28)$$

$$R = -2 \sum_{j=N+1}^{N+M} e(j/\underline{\theta}) \, \frac{\partial e(j/\underline{\theta})}{\partial \underline{\theta}} \quad (29)$$

and $\dfrac{1}{((N+iM)/(M+1))}$ is the stochastic gain.

The extended Kalman Filter approach can be used for parameter estimation by employing an augmenting of the "state vector" (to be estimated) to include not only the system states, but also the parameters to be identified. Given the system

$$\underline{x}(k+1) = A\,\underline{x}(k) + B\,u(k) + d\underline{w}(k) \qquad (30)$$

$$z(k+1) = h\underline{x}(k+1) + v(k+1)$$

then an augmented state vector is

$$\underline{X}^T(k) = [\underline{x}^T(k),\, \underline{a}^T,\, \underline{b}^T] \qquad (31)$$

with $A = \begin{bmatrix} 0 & & I & \\ \hline a_1 & a_2 & \cdots & a_n \end{bmatrix}$ $B^T = [b_1,\, b_2\, \ldots b_n]$, $h = [1\,0\,0\ldots 0]$,

$d = [d_1,\, d_2,\, \ldots d_n]\, u(k)$ a scalar, and $\underline{w}(k)$ and $v(k)$ are zero mean Gaussian random vectors with convariances Q, R. It is assumed $\underline{a}$ and $\underline{b}$ are constant, or

$$\underline{a}(k+1) = \underline{a}(k),\ \underline{b}(k+1) = \underline{b}(k) \qquad (32)$$

The augmented system equations are

$$\underline{X}(k+1) = \underline{F}\bigl(\underline{X}(k),\, u(k)\bigr) + d_1 \underline{w}(k) \qquad (33)$$

$$z(k) = H^T \underline{X}(k) + v(k) \qquad (34)$$

where H, d_1 are 3n-dimensional vectors and $\underline{F}$ is a nonlinear function vector. Expanding (33) in a Taylor Series, one obtains

$$\underline{F}\,(\underline{X}(k),\, \underline{u}(k)) = \phi(k)\underline{X}(k) + \text{h.o.t.} \qquad (35)$$

The term ϕ is of the form

$$\phi(k) = \begin{bmatrix} 0 & I & 0 & \\ \hline \underline{a}(k) & & \hat{\underline{x}}(k) & u(k)I \\ \hline 0 & & I & 0 \\ \hline 0 & & 0 & I \end{bmatrix} \qquad (36)$$

$$\hat{\underline{x}}(k+1/k) = \underline{F}(\hat{\underline{x}}(k/k), u(k)) \tag{37}$$

$$P(k+1/k) = \phi(k) P(k/k) \phi^T(k) + Q d_1 d_1^T \tag{38}$$

$$P(k+1/k+1) = P(k+1/k) - P(k+1/k) H$$
$$[H^T P(k+1/k) H + R]^{-1} H^T P(k+1/k) \tag{39}$$

The previous methods discussed have been techniques for the identification of parameters for dynamic systems. But there are situations of interest in power where static system parameter estimation is of value [42], such as in security monitoring, wherein one checks the steady-state operation of a system on-line. The static estimation means that the significant transient effects have died out. Although similar in form to dynamic parameter identification, algorithms for static estimation assume a simpler form due to algebraic expressions (as opposed to dynamic equations) being used.

Assuming a system in steady-state with constant parameter vector $\underline{p}$, the measurement equation vector can be written

$$\underline{z}(k) = \underline{h}(\underline{x}(k),\underline{p}) + \underline{v}(k) \tag{40}$$

where $\underline{h}$ is the nonlinear transformation, $\underline{x}(k)$ system state vector, and $\underline{v}$ a noise. To use a variation on the extended Kalman filter approach discussed previously, define

$$\underline{x}(k+1) = \underline{x}(k) + \underline{w}(k) \tag{41}$$

$$\underline{p}(k+1) = \underline{p}(k) \tag{42}$$

where $\underline{w}(k)_1$ is a zero mean gaussian random vector with covariance Q (Q^{-1} assumed zero here). The updating algorithm for the parameter $\underline{p}$ at time $t = k$ is [43]

$$\Delta\hat{\underline{\theta}}(i+1) = P(i) \begin{bmatrix} H_x^T R^{-1} \underline{z}(i) \\ \hline H_p^T R^{-1} \tilde{\underline{z}} + M^{-1} \tilde{\underline{p}}(i) \end{bmatrix} \tag{43}$$

where $\Delta\hat{\underline{\theta}}^T = [(\hat{\underline{x}}(i+1) - \hat{\underline{x}}(i))\ (\hat{\underline{p}}(i+1) - \hat{\underline{p}}(i))]$, i is the iteration counter (how many times an estimate at k occurs),

$$H_{\underline{x}} = \frac{\partial \underline{h}}{\partial \underline{x}}\bigg|_{\hat{\underline{x}}(i)}, \quad H_p = \frac{\partial \underline{h}}{\partial \underline{p}}\bigg|_{\hat{\underline{p}}(i)},$$

$$P(i) = \begin{bmatrix} P_{xx}(i) & | & P_{xp}(i) \\ \hline P_{px}(i) & | & P_{pp}(i) \end{bmatrix}^{-1} \quad (44)$$

$$P_{xx} = H_{\underline{x}}^T R^{-1} H_{\underline{x}}$$

$$P_{xp} = H_{\underline{x}}^T R^{-1} H_{\underline{p}}$$

$$P_{px} = H_{\underline{p}}^T R^{-1} H_{\underline{x}} \quad (45)$$

$$P_{pp} = M^{-1} + H_{\underline{p}}^T F^{-1} H_{\underline{p}}$$

$$M = P_{pp}\bigg|_{t=k-1}$$

$$\underline{z} = \underline{z} - \underline{h}(\hat{x}(i), \hat{p}(i)) \quad (46)$$

$$\underline{p}(i) = \hat{\underline{p}}\bigg|_{t=k-1} - \hat{\underline{p}}(i)\bigg|_{t=k} \quad (47)$$

and initial condition $p(0)\bigg|_{t=k} = p\bigg|_{k-1}$

V. POWER SYSTEM IDENTIFICATION EXAMPLES

A. Large-Scale System Dynamic Equivalent

To demonstrate the ARMA least-square identification approach, consider the 345 KV New England system shown in Figure 7. It has 39 busses, 46 lines and transformers, and 10 generators, all of which interact. It is desired to determine a series of "best" fit low-order models of the type given in (10), with which accurate security assessment studies can be undertaken. By hypothesizing various order and type

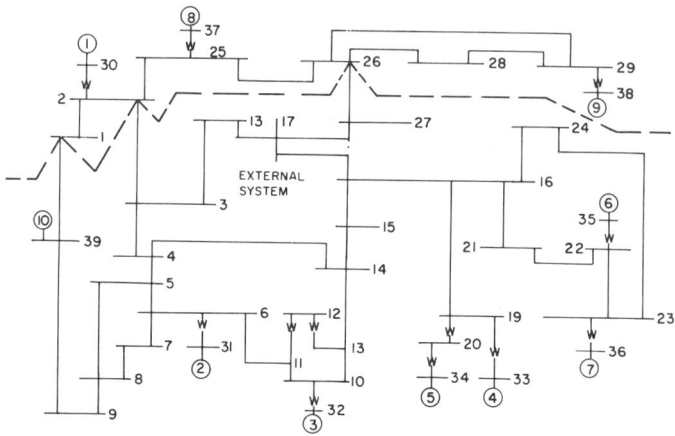

Fig. 7. Test System for Dynamic Equivalents Example

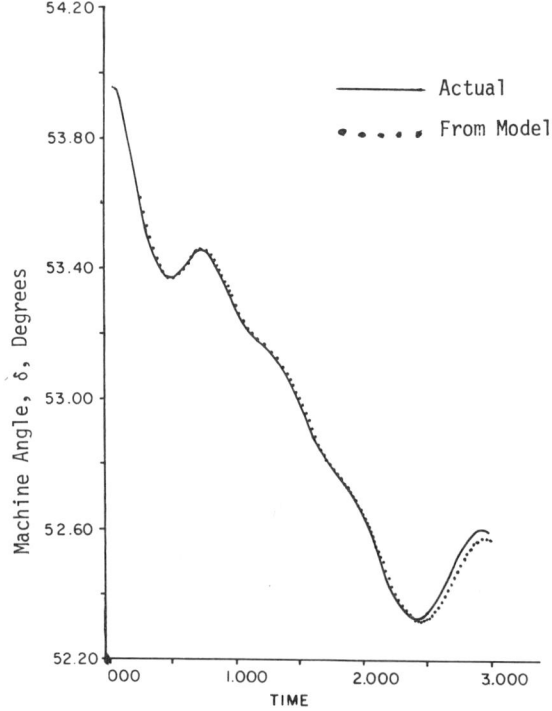

Fig. 8. Typical Machine Swing Curve; Model vs. Actual

models (as given in [3]), a determination of the best models can be made. This task was performed and a series of dynamic equivalents developed. A time-domain simulation of the actual system vs. the models was run and some of the results are given in Figure 8. Here one of the system generator swing curves is given. Note the excellent correlation between the two curves, suggesting that it is possible to obtain meaningful low-order fits.

B. Frequency Prediction From Tie Line Measurements [4]

A linear, time-varying model relating frequency f and power p is assumed as

$$f(i) = [\sum_{j=1}^{n} a_{ij}f(i-j) + b_{ij}p(i-j)] + e(i) \quad (48)$$

where e is noise representing modeling errors, and a_{ij}, b_{ij} depend on inertia and damping of generators and synchronizing torque coefficients of tie lines electrically near measurement location, and n is the system order. Using on-line least squares methods on a 60 Hz. line predicting ahead in time the expected line frequency, results in a prediction frequency error as shown in Figure 9, where the y axis is

$$\frac{1}{N} \sum_{i=1}^{N} (f(i) - \hat{f}(i))^2 \quad (49)$$

where N is the number of data points employed. Typically, results indicate frequency can be predicted up to one second in advance.

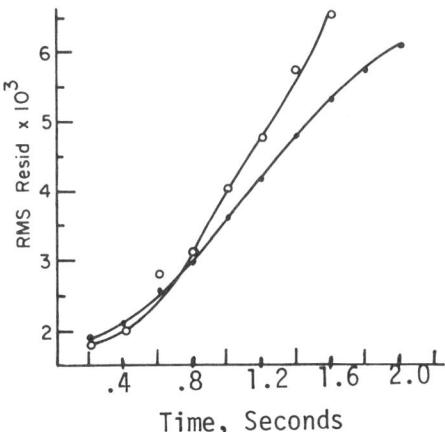

Fig. 9. Frequency Prediction Error vs. Time Advance

C. Model-Reference Identification

A power system has the following equivalent structure

$$\dot{\underline{x}} = \begin{bmatrix} a_1 & 1 \\ a_2 & 0 \end{bmatrix} \underline{x} + \begin{bmatrix} b_1 \\ b_2 \end{bmatrix} u \qquad (50)$$

$$y = x_1 = \text{output}$$

and it is desired to determine the constant parameters a_1, a_2, b_1, b_2 only from operating data y and u. Using an MRAS observer approach developed in [44] with $a_1 = 5$, $a_2 = -10$, $b_1 = 1$, $b_2 = 2$ and u = periodic rectangular pulses of height 5 and frequency 1 rad/sec, system identification was effected. Some of the results are given in Figure 10. Note the relatively rapid convergence of the parameter estimates to the correct values. Also note that a continuous parameter estimate is available for on-line use in control, if required [45].

D. Regular Tracking Problem

Identification can be of value in power system control in the case of optimal controllers [46]. Most theory for optimal control requires a knowledge of the plant dynamics, which are not known in many instances. As shown in Figure 11, identification can be used to determine the necessary parameters for the controller, so as to regulate such quantities as frequency, voltage, MVARS and MW, current, etc. A variation on this idea is to develop a control law directly from measurements of inputs and outputs, without explicitly identifying the plant parameters [47].

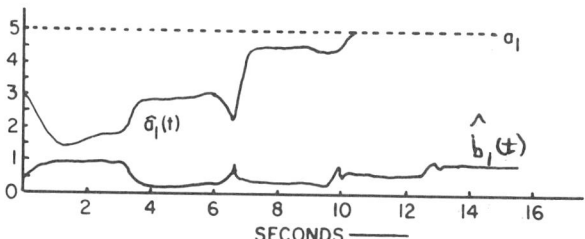

Fig. 10. MRAS Identification Response Results

REGULATOR TRACKING PROBLEM:

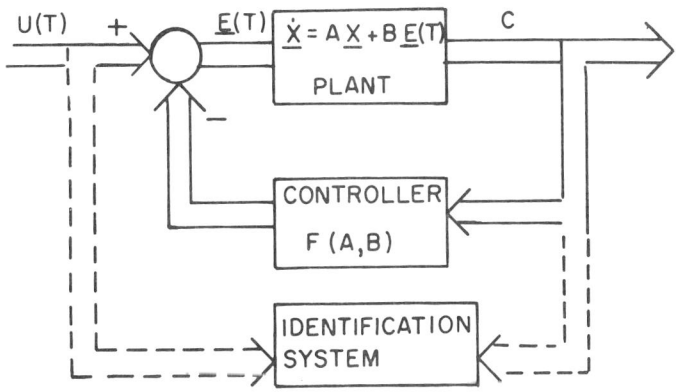

Fig. 11. A Regulator Tracking Approach Using Identification

VI. CONCLUDING COMMENTS

The applications of systems identification techniques in power systems are many. A vast amount of literature has been written concerning different identification forms, although exhaustive comparisons of methods are few. One of the difficulties of practically applying these and other techniques is an often-required restriction on the nature of the input, the so-called "frequency rich" or "sufficiently exciting" condition [25, 48]. The difficulty is that these theoretical constraints are not always met by power signals without deliberate perturbation of the power drive signal, a questionable situation at best.

Another area of concern with power system identification which has not been fully developed yet is that of the order, structure, and nature of the models used. Although much work has been done from systems theory viewpoint [49 – 51], more applied development work is needed.

VII. REFERENCES

(1). Gaus, K.G., Theory of Motion of the Heavenly Bodies, New York: Dover, 1963.
(2). Rubenstein, E., "Cash Crunch Dims Power Prospects," IEEE Spectrum, March 1975, pp. 40-44.
(3). Ibrahim, M., Mostafa, O. and El-Abiad, A., "Dynamic Equivalents Using Operating Data and Stochastic Modeling," IEEE Trans. PAS, vol. PAS-95, no. 5, September/October 1976.

(4). Schwartz, M., et. al., "Model Identification Using Tie Line Power and Frequency Measurements," Automatica, vol. 12, no. 1, January 1976, pp. 23-30.

(5). Horisberger, H., et. al., "A Fast Decoupled Static State Estimator For Electric Power Systems," IEEE Trans. PAS, vol. PAS-95, no. 1, January/February 1976, pp. 208-215.

(6). Taylor, C. and Cresap, R., "Real-Time Power System Simulation For Automatic Generation Control," IEEE Trans. PAS, vol. PAS-95, no. 1, January/February 1976, pp. 375-384.

(7). Happ, H., et. al., "Large-Scale Hydro-Thermal Unit Commitment," PICA Conference Proceedings, Denver, May 1969.

(8). Prabhakara, F. and El-Abiad, A., "A Simplified Determination of Transient Stability Regions For Lyapunov Methods," IEEE Trans. PAS, vol. PAS-94, no. 2, March/April 1975.

(9). Thomas, R. and Thorp, J., "A Model-Referenced Controller for Stabilizing Large Transient Swings in Power Systems," IEEE Trans. Auto. Control, vol. AC-21, no. 5, October 1976, pp. 746-750.

(10). Dash, P., et. al., "Transient Stability and Optimal Control of Parallel AC-DC Power Systems," IEEE Trans. PAS, vol. PAS-95, no. 3, May/June 1976, pp. 811-820.

(11). Kumar, A. and Richards, E., "A Suboptimal Control Law To Improve The Transient Stability of Power Systems," IEEE Trans. PAS, vol. PAS-95, no. 1, January/February 1976, pp. 243-247.

(12). Kwaty, H., et. al., "An Optimal Approach To Load-Frequency Control," IEEE Trans. PAS, vol. PAS-94, no. 5, September/October 1975, pp. 1635-1643.

(13). Cavin, R., "An Optimal Linear Systems Approach To Load-Frequency Control," IEEE Trans. PAS, vol. PAS-90, no. 6, December 1971, pp. 2472-2482.

(14). Masiello, R. and Schweppe, F., "Multi-Machine Excitation Stabilization Via System Identification," IEEE Trans. PAS, vol. PAS-94, no. 2, March/April 1975, pp. 444-454.

(15). Weiss, "Transient Asymptotic Stability of Power Systems as Established With Lyapunov Functions," IEEE Trans. PAS, vol. PAS-95, no. 4, July/August 1976.

(16). Webb, R., Womble, M., et. al., "Alternator Modeling, Part I. Simulation, Parameter Estimation, and Output Sensitivity Analysis," Technical Report AFAPL-TR-75-87, School of Electrical Engineering, Georgia Institute of Technology, October 1975.

(17). Happ, H., "Optimal Power Dispatch - A Comprehensive Survey," IEEE Trans. PAS, vol. PAS-96, no. 3, May/June 1977, pp. 841-854.

(18). Ward, J.W., "Equivalent Circuits for Power Flow Studies," Trans. AIEE, vol. 68, 1959, pp. 373-383.
(19). Zadeh, L.A., "From Circuit Theory to System Theory," Proc. IRE, vol. 50, pp. 856-865, 1962.
(20). Galiana, D., et. al., "Identification of Stochastic Electric Load Models From Physical Data," IEEE Trans. Auto. Control, vol. AC-19, December 1974, pp. 887-892.
(21). Quan, R., and Tarnawecky, M., "Load Representation for Transient Stability Studies-Digital Modeling," Paper presented at IEEE PES Summer Meeting, 1975.
(22). Aström, K. and Eykhoff, P., "System Identification - A Survey," Automatica, vol. 7, pp. 123-162, 1971.
(23). Clarke, D., "Generalized-Least-Squares Estimation of the Parameters of a Dynamic Model," IFAC Symp. Identification, 1967.
(24). Kashyap, R., "Maximum Likelihood Identification of Stochastic Linear Systems," IEEE Trans. Auto. Control, vol. AC-15, no. 1, 1970, pp. 25-34.
(25). Lüders, G. and Narendra, K.S., "An Adaptive Observer and Identifier for a Linear System," IEEE Trans. Auto Control, vol. AC-19, October 1973, pp. 496-499.
(26). Landau, I.D., "Unbiased Recursive Identification Using Model Reference Adaptive Techniques," IEEE Trans. Auto. Control, April 1976, pp. 194-202.
(27). Colburn, B.K. and Boland, J.S., "Analysis of Error Convergence Rate for a Discrete Model-Reference Adaptive System," IEEE Trans. Auto Control, vol. AC-21, no. 5, October 1976.
(28). Hang, C., "A New Form of Stable Adaptive Observer," IEEE Trans. Auto. Control, vol. AC-21, no. 4, August 1976, pp. 544-547.
(29). Young, P., "An Instrumental Variable Method for Real-Time Identification of a Noisy Process," Automatica, vol. 6, 1970, pp. 271-287.
(30). Hill, J. and McMurtry, G., "An Application of Digital Computers to Linear System Identification," IEEE Trans. Auto. Cont., vol. AC-9, 1964, pp. 536-538.
(31). Mehra, R., "On-Line Identification of Linear Dynamic Systems With Applications to Kalman Filtering," IEEE Trans. Auto. Cont., April 1970, pp. 175-184.
(32). Brogan, W.L., Modern Control Theory, Quantum Publishers: New York, 1974, ch. 17.
(33). Ästrom, K., "On the Achievable Accuracy in Identification Problems," Proc. IFAC Symposium on Identification in Automatic Control, Paper 1-8, June 1967.
(34). Box, G. and Jenkins, G., Time Series Analysis Forecasting and Control, Holden Day: San Francisco, 1970, Ch. 3.

(35). Hastings, James, R. and Sap, M., "Recursive Generalized Least-Squares Procedure for On-Line Identification of Process Parameters, Proc. IEEE, vol. 116, 1969, pp. 2057-2062.

(36). Saridis, G., "Comparison of Six On-Line Identification Algorithms," Automatica, vol. 10, no. 1, 1974, pp. 69-79.

(37). Isermann, R., et. al., "Comparison of Six On-Line Identification and Parameter Estimation Methods," Automatica, vol. 10, no. 1, 1974, pp. 81-103.

(38). Gustavsson, J., "Comparison of Different Methods for Identification of Industrial Processes," Automatica, vol. 8, 1972, pp. 127-142.

(39). Landau, I.D., "A Survey of Model-Reference Adaptive Techniques - Theory and Applications," Automatica, vol. 10, 1974, pp. 353-379.

(40). Eykhoff, P., System Identification: Parameter and State Estimation, Wiley-Interscience: New York, 1974.

(41). Nuhan, S. and Carroll, R., "An Adaptive Observer and Identifier With An Arbitrarily Fast Rate of Convergence," Proc. of 1976 Conf. Decision Control, pp. 1081-1087.

(42). Kim, C., and Lindorff, D., "Generalization of Adaptive Observer and Method of Optimizing its Convergence Properties," ASME J. Dyn. Sys., Meas., Control, March 1977, pp. 15-22.

(43). Debs, A., "Parameter Estimation for Power Systems in the steady state," IEEE Trans. Auto. Cont., vol. AC-19, no. 6, December 1974, pp. 882-886.

(44). Kudva, P. and Narendra, K.S., "Synthesis of an Adaptive Observer Using Lyapunov's Direct Method," Int. Journal of Control, December 1973, pp. 1201-1210.

(45). Narendra, K.S., "Stable Identification Schemes," in System Identification: Advances and Case Studies, edited by R. Mehra and D. Lainiotis, Academic Press, 1976.

(46). Wilson, W., et. al., "Nonlinear Output Feedback Excitation Controller Design Based on Nonlinear Optimal Control and Identification Methods," Paper No. A 76 343-4, IEEE Summer Power Meeting, 1976.

(47). Walker, P. and Dunne, M., "Direct Formation of a Time-Optimal Controller by Least-Squares Identification," IEEE Trans. Auto. Cont., vol. AC-22, no. 4, August 1977, pp. 669-670.

(48). Mehra, R., "Frequency-Domain Synthesis of Optimal Inputs for Linear System Parameter Estimation," ASME J. Dynamic Systems, Measurements, and Control, June 1976, pp. 130-138.

(49). Woodside, C., "Estimation of the Order of Linear Systems," Automatica, vol. 7, 1971, pp. 727-733.

(50). Unbeauen, H. and Gohring, B., "Tests for Determining Model Order in Parameter Estimation," Automatica, vol. 10, 1974, pp. 245-256.
(51). Van Den Boom, eta. al., "The Determination of the Orders of Process and Noise Dynamics," Automatica, vol. 10, 1974, pp. 245-256.

Acknowledgement

I would like to thank Dr. O. Mostafa for his assistance with the section on dynamic equivalents, and Mr. K. R. Chakravarthi, Power Engineer, Southern Company Services, Birmingham, Alabama for his counsel regarding power system applications and examples.

Power Control System and Protection

Synchronous Machine Modeling
A System Comparison

John E. Fagan
The University of Oklahoma
Norman, Oklahoma

BACKGROUND

Historically, synchronous machine model developments have been primarily confined to the modeling of only the machine and have usually been of a theoretical nature. Consequently, the literature has produced overly complex models of the synchronous machine (1, 2, 3, 4) which are largely used for the simulation of the steady state or dynamic performance of the individual machine. The large amount of data required and the complexity of the models make them undersirable for system level studies involving many machines tied by a network configuration. The size of the system of equations to be solved does not lend itself to a solution using today's numerical methods and digital hardware. Thus, a simpler modeling approach is desirable.

MACHINE MODELING

In order to facilitate the solution of even small systems of machines (2 or 3), R.H. Park in 1929 proposed a simplified representation of the salient pole synchronous machine. (37) Later, Charles Concordia (6, 7) expanded the Park model to include the solid rotor as well as salient pole machines. The Park model was suitable for the limited computational tools available at that time such as analog computers and limited network analyzers. However, for power systems stability analysis it was soon necessary to model the behavior of many interconnected machines. Because of the demands, a number of investigations were conducted in order to define an adequate simplified form of the Park equations. These investigations formed the basis for present models used in power system stability analysis. (8,3 9, 10,11)

It has been shown that the interaction of the synchronous machine and its controller systems has a direct effect on the simplifications that can be made in machine representation applied to large scale system studies. (12) Based on the types of controllers (exciters and governors) that were prevalent in that period of time, the "classical" representation of the synchronous machine evolved. (10) System operating margins gradually diminished due to economic and environmental restrictions, the use of high speed excitation systems, and the addition of such controllers as power system stabilizers and governor fast valving. Thus, the need to represent the generator system in more detail for transient, as well as dynamic, system studies became imperative.

SYSTEM SIMULATION

Recently, a few commercial programs were made available to solve the power system transient stability problem.(39, 51) These programs generally have a limited number of machine representations available and in some cases the machine model is fixed to a single representation.

Wide use of these digital simulators by the industry has caused much confusion about the often drastically different system results obtained using different system simulators to model the same system. This confusion has led to a distrust of some of the digital simulators and a lack of understanding of simulation results.

Research has revealed that the misunderstanding could be removed by consideration of the impact of the various machine models used in these system level studies.

The machine model comparison shown here reveals the current disparity between today's commercial system simulators. It clearly, shows the impact of the machine model on power systems dynamic studies. It also graphically shows the importance of the machine model to studies which determine the bus critical fault clearing time used for selection of system breaker and relay protection equipment. The comparison puts into proper perspective the relative importance of machine and controller representation for system simulation studies. The comparisons are important to the power industry in understanding the role of the machine model in power system dynamic simulation studies. They are also important since a proper understanding of machine model use, and performance is necessary to the development of a load model compatible with existing power system dynamic solution techniques.

STUDY SYSTEM

In order to observe the different levels of machine models in a system situation, a sample system was chosen. The 10 machines in the system were the only variables in the system to change. Each of the 10 machines were equipped with an IEEE type 1 excitation system(15) and a Philadelphia Electric Co. (PECO) Generalized Steam and Hydro model.(13) The system used for the study is shown in Figure 1. It is the EPRI 39 bus, 46 line, 10 generator representation of the New England bulk transmission system.(16) The data has been expanded from the EPRI data in order to be able to support the higher level models. The data was expanded based on the individual characteristics of each machine.

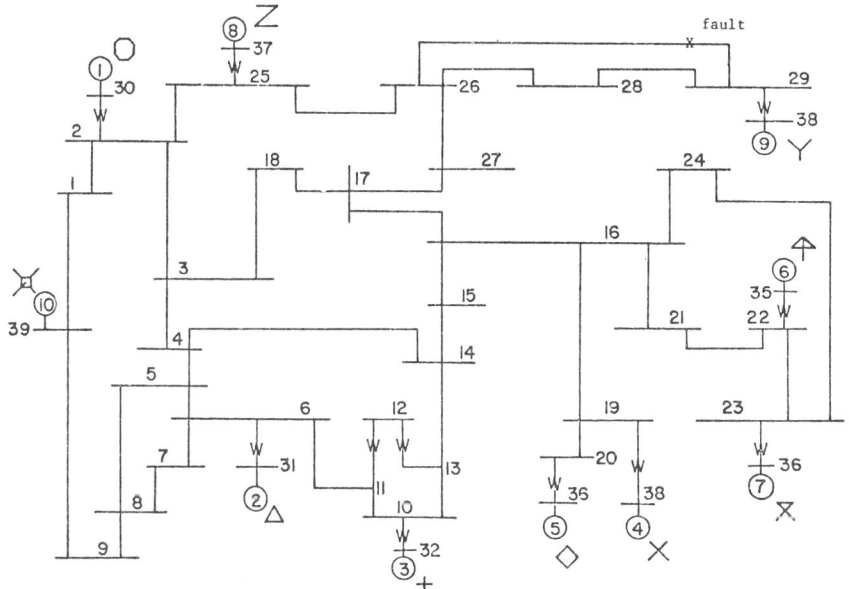

Figure 1 Sample EPRI 39 Bus 10 Generator Test System

For the majority of the studies shown the sample system was subjected to a 4.5 cycle (0.75 sec) three phase fault on the 26-29 line side of bus 29, the fault location is shown in Figure 1. The location and level of the disturbance was based on the following:

1. The fault was placed on the high side of the machine transformer, slightly distant from the machine, thus avoiding the level of d-c components that would be established by a closer fault to machine 9.
2. Machine 9 is loosely coupled to the system through the lines lines (26-28), (28-29) and (26, 29). Hence, a fault at bus 29 would serve to exhibit the model test under severe and small disturbance conditions, bus 9 being severely disturbed and the remainder of the system buses having only moderate disturbance.
3. The clearing action, removal of line 26-29, further serves to loosely couple the machine 9 to the rest of the system.

MODEL COMPARISON

In order to more graphically see these model differences, a comparison is made between the individual model system studies. The relative effects of machine modeling on a single pair of machines will be individually compared from each of the previously conducted system studies. The machine of highest interest to the power systems engineer is machine 9 of Figure 1 since it is the unit most effected by the system disturbance and, therefore, useful to a comparative study of the effects of modeling in close proximity to a severe system disturbance. Additionally, a machine representative of the coherent group will also be comparatively studied. This machine is only moderately effected by the stystem disturbance, thus enabling the reader to see the effects of machine modeling under conditions of both moderate and severe system disturbance levels.

Initially a comparison is made between the transient rotor angle profiles of machine 9 under 4 levels of model. This comparison is shown in Figure 2. Model 1 (17) is not shown since it exhibited an unstable response under the system disturbance used to test the higher level models. Model 5 (17) produced the least pessimistic system results of any of the models used. However, the degree of improvements from model 3 to 5 is not as pronounced from model 3 to 4 nor 2 to 3. With the tighter system response exhibited by the higher level models is seen also an increase in the characteristic frequency exhibited by the system response. Comparison can also be seen in Figure 2, a comparison of the rotor angle response of machine 9 under varying machine models. Figure 3 is a comparison of machine 9's dynamic voltage profile under the 4 higher levels of machine model. The dynamic voltage profile is improved with each level of model while, as shown in Figure 4 the machine exciter response to the system disturbance is progressively smaller. This is a clear indication of the degree of damping provided by the rotor as compared to that provided by the excitation system. As the model is improved by increased rotor representation, the role of the excitation system is diminished in the system study. These results indicate that the model performance gain by the additional rotor circuit representation is substantial over the excitation system model improvements. The payoff for the excitation system is in the long term response within an individual model's levels.

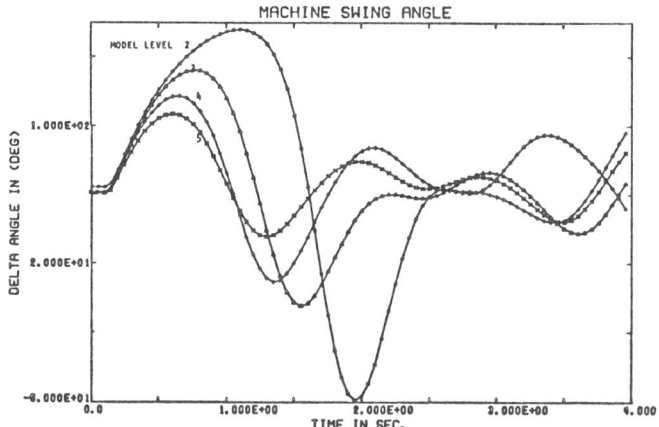

Figure 2 Machine 9 Transient Rotor Angle Comparison

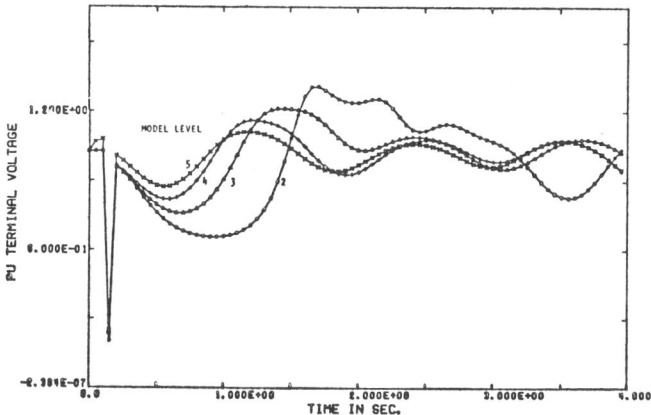

Figure 3 Machine 9 dynamic voltage profile.

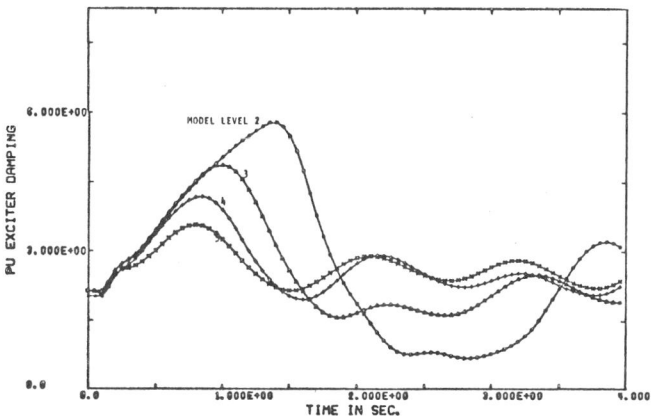

Figure 4 Machine 9 Execitation System Response.

The comparative study of unit 9 indicates that the machine model has a significant effect on the overall performance of a power system simulation when the unit is subjected to a large disturbance. This is a critical fact since the prime interest in a system simulation stems around the machine most effected by the system disturbance. One question to be looked at in a later section is that at each level of model improvement there was a change in system response. This response change became smaller at each higher level of model. So the question asked is at what level of model do the results cease to change? Before this can be examined, however, a similar model performance comparison must be made on a machine not subject to a major system disturbance. For this comparison a machine representation of the coherent group of machines was chosen since these machines were only subject to a minor disturbance by the fault action at bus 29. The machine chosen is unit 6. Figure 5 presents a transient rotor angle comparison of unit 6 under the 4 highest levels of machine model. The differences in machine modeling are most subtle when the machine is subjected to only moderate system disturbances. The greatest change in performance is seen in going from model 2 to 3. The changes seen in going to the still higher level models are quite subtle; the performance of model 3, 4 and 5 are almost indentical. This would indicate that the further the machine is removed from the system disturbance, the less important the modeling is. This comparison might also indicate that in this case machine model 3 would be the minimal model that the system simulator would use. It should be noted here that the performance of unit 6 is typical of that seen of the other machines in the system. Only unit 6 was presented since the performance of the other machines would give only redundant results.

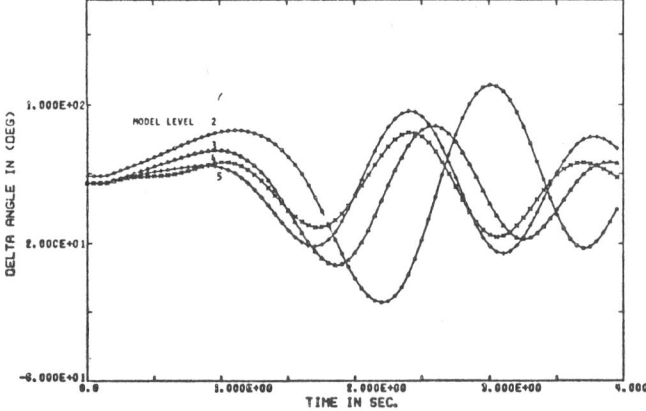

Figure 6 Machine 6 Transient Rotor Angle Comparison Unit the Five levels of machine models.

There were no significant differences in the machine's voltage profile under the different machine models and, hence, the graphic results are not presented.

A COMMERCIAL SIMULATION COMPARISON

For many years the utility industry has relied on several commercially available dynamic simulation programs for system planning purposes. Three of these programs are the Philadelphia Electric Co (PECO) Program, the Westinghouse Power System Stability Program (WPSS) and the University Computer Services Stability Program (UCS), to name a few. Both the UCS and WPSS programs are available to the utility industry system planner on a cost per run basis thru the vendor's computer, or the WPSS program can be leased to the user installation. The PECO program is free for a one time reproduction charge and can be easily installed on a utility's IBM 360 or 370 series of computers having at least 256K bytes of user core and disk storage of the 3330 type. Due to the cost involved the PECO program is the most widely used program by the utility industry.

For many years the utility planners have obtained often drastically different results from the three system simulator programs for the same system studies. These different results have led to confusion in the industry as to how to interpret the results from the individual simulators. Many efforts have been made to explain the simulator differences from integrator problems to controller representation. The machine model was largely ignored since little was known about its system behavior together with the exact form the models took within the programs. It was also felt that this did not contribute substantially to the overall system results.

During this research the reason for the programs' differences were found, in fact, to be due only to the machine models each simulator uses. The performance of each program is directly correlated to machine modeling used.

A study was conducted to compare the results obtained with the PECO, USC, and Westinghouse programs on a single small test system. This test was conducted by a local utility using a modified version of the 39 bus system used in this research to demonstrate the performance of each model. The network configuration was identical and differed only slightly in 4 of the machine models. A .065 sec. fault was placed on bus 29 as in the model tests and the fault was cleared by removing line 26-29. The study was only one second in duration since this is about the stability limits of the PECO program. The results are shown in Figure 7 thru 10. As in the earlier test system generator 9 was the machine most affected by the system disturbance, hence, only the results of this machine were looked at. The PECO, UCS, and Westinghouse results for generator 9 are plotted together with the author's results for model 2 and model 3 machine representation. The PECO results almost agree completely with the model 2 results, which is what would be expected since the PECO program uses a form of level 2 modeling. The UCS and Westinghouse studies agree with model 3 results since the data provided these programs forced the program to choose a form of model 3 representation. Figure 6 is a comparison of the program's transient rotor angle plots for machine 9. Figures 7, 8 and 9 compare significant state variables of the machine 9. In all comparisons the model 2 compares favorably with the PECO representation while the model 3 results compare with the USC and Westinghouse studies. Additional studies that were run arrived at the same general comparisons.

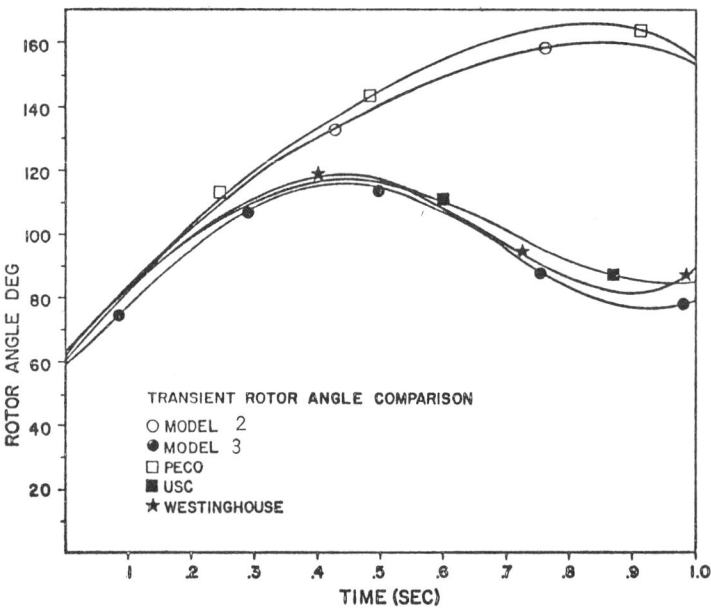

Figure 7

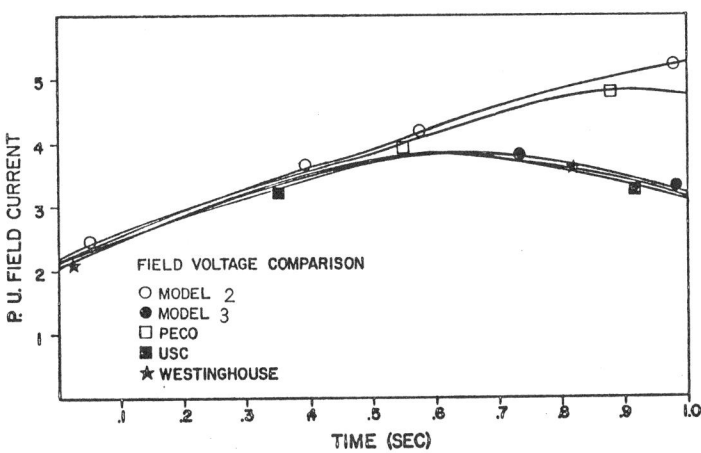

Figure 8

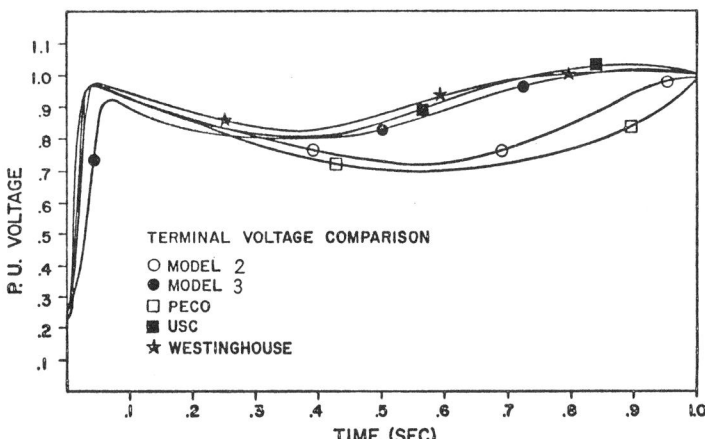

Figure 9

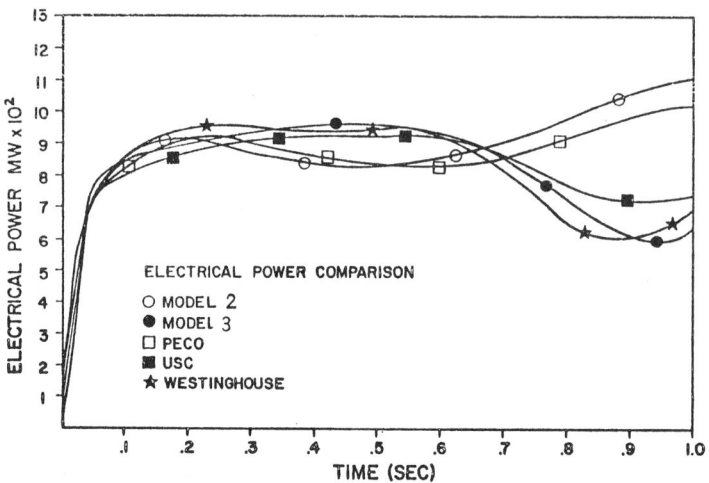

Figure 10

The comparisons made in Figures 7 and 10 graphically explain the program's differences in system results. The program's results are directly related to the machine model chosen since exciters and governors were the same for the studies.

The PECO program is not full of errors as is suspected by some of the industry but is giving results consistant with the model implemented within. Also, the Westinghouse programs are producing results consistant with <u>user</u> selected models. For these reasons results gained from the simulators should be interpreted for the first time in light of the machine models being used during the simulations.

MODEL CRITICAL FAULT CLEARING TIME

During system planning, breakers and relaying for the purposes of system protection are chosen based largely on stability studies to determine the system's critical fault clearing times. These studies have a large impact on system protection costing in that if the stability studies indicate the system can remain in tact, and if the system faults are cleared within specific times, then breaking equipment is purchased according to its speed.

FAULT CLEARING TIME (hertz)

SYSTEM Disturbance	MODEL 1	MODEL 2	MODEL 3	MODEL 4	MODEL 5
30 Fault 29 [1]Clear 26-29	2.40	4.80	5.40	5.80	6.60
30 Fault 25 [2]Clear 2-25	2.60	5.70	6.60	7.0	7.5
30 Fault 17 [3]Clear Remove Fault	6.4	13.8	14.2	14.5	14.5

TABLE 4-1

The machine model has a profound effect on the critical fault clearing time, studies will be shown. Three different system contingencies were run on the sample system of Figure 2. The studies were run until the particular critical fault clearing time was determined for the 5 levels of machine models. These results are shown in Table 1.

In each case the system fault clearing time increased with model level. Where the fault is close to the machine and the clearing action resulted in a weak tie to the rest of the system, disturbance 1 and 2, the machine model made a large difference in the buses' critical fault clearing time for that particular disturbance. As the disturbance was moved to a tight group of machines, disturbance 3, the model difference produced less drastic differences in fault clearing times between the representations 2, 3, 4 and 5, however, model 1 had a significantly lower critical fault clearing time for this disturbance.

The impact of this model comparison is quite profound. If the system planner is using largely model 2 and 2 machine representations, which is the case in today's studies due to use of the PECO simulator, then the system protection installed may not be the most cost effective.

MODEL COMPARISON UNDER LONG TERM STUDIES

As part of the model comparison made was, the effect of machine modeling and dynamic stability studies. Currently, it is the practice to use extremely simple machine models 1 for dynamic system simulations. These simple models are equipped with linear governor-turbine representations but no exciter representations. These simplified and linearized machine system representations are used to effect a fast solution to the dynamic simulation problem. These dynamic simulators are used only for the study of small system disturbances where the frequency deviation is usually on one or two hertz. However, dynamic stability problems are in some cases due to the after effects of a major system disturbance and these simulators cannot model this type of system event.

With this in mind a group of dynamic stability runs was made using the program contained herein. The effects of machine modeling were compared on dynamic simulations. Each machine was equipped with an exciter and governor-turbine system as in earlier system tests. The system was then disturbed using the 3 phase .075 sec. fault on bus 29 and clearing the fault by removing line 26-29. The study was conducted for a total of 16 seconds of system simulation. A comparison of the dynamic rotor angle plots for generator 9 was made and is shown as Figure 11. Representations of model levels 2, 3 4 and 5 were compared. These models were 2, 3, 4 and 5 respectively.

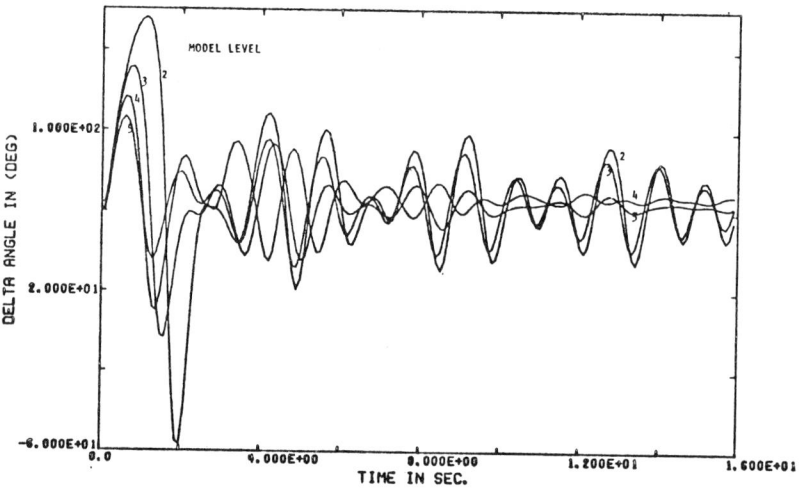

Figure 11 Generator a model comparison for long term study.

The higher level models exhibited better first swing characteristics from 0-3 seconds of simulation time. From 3 to 7 seconds all models behaved about the same. However, beyond 7 seconds the higher model representations showed much better and more realistic recovery than the two simpler models. This comparison graphically points out that machine modeling has a distinct effect even on long term stability studies where historically, the machine model has been highly simplified in order to achieve a fast solution.

SUMMARY

This paper has presented a graphic comparison of the effects of machine models on power system dynamic simulations. The effects of the models on transient as well as dynamic stability studies has been shown, together with the models' impact on critical fault clearing time studies. The differences between major system simulation packages have been explained as a machine model rather than solution process or controller problem. The paper detailed the sensitivity of the power system simulation to the machine models used to represent the system synchronous machines.

BIBLIOGRAPHY

(1) Grigsby, L. L., "Transient and Dynamic Stability -- Past, Present and Future," <u>Control of Power System Conference Proceedings</u>, 76CH1057-0, 1976.

(2) Subramaniam, P., and Malik, O. P., "Digital Simulation of a Synchronous Generation in Direct Phase Quantities," <u>Proc. IEEE</u>, Vol. 118, No. 1, 1971, pp. 153-160.

(3) Dineley, J. L. and Morris, A. J., "Synchronous Generator Transient Control Part 1--Theory and Evaluation of Alternative Mathematical Models," <u>IEEE/PAS</u>, Vol. PAS-92, Mar/Apr. 1973, pp. 417-422.

(4) Rankin, A. W., "The Equations of the Idealized Synchronous Machine," <u>General Electric Review</u>, June 1944, pp. 31-36.

(5) Park, R. H., "Two-Reaction Theory of Synchronous Machines, Part 2," <u>AIEE Trans.</u>, Vol. 52, June 1933, pp. 353-355.

(6) Concordia, C. and Poritsky, H., "Synchronous Machine with Solid Cylindrical Rotor," <u>AIEE Trans.</u>, Vol. 59, 1937, p. 49.

(7) Concordia, Charles, <u>Synchronous Machines Theory and Performance</u>, General Electric Press, 1951.

(8) Crary, S. B., <u>Power System Stability</u> (in two volumes), New York: John Wiley and Son, Inc., 1945.

(9) Kimbark, E. W., <u>Power System Stability</u>, (in 3 volumes), New York: John Wiley and Son, 1956.

(10) AIEE Subcommittee on Interconnection and Stability Factors, "First Report of Power System Stability," <u>AIEE Trans</u> Vol. 56, 1937.

(11) IEEE Working Group on Power System Stability, "Power System Stability Bibliography 1935-1965," <u>IEEE Special Publication 31</u>, pp. 67-192.

(12) Young, C. C., "The Synchronous Machine," <u>Modern Concepts in Power Systems Dynamics</u>, IEEE Press, 70M62-PWR, 1970.

(13) "Philadelphia Electric Co. Stability Program Users Guide," <u>PECO</u>, 1976.

(14) "Westinghouse Co. Stability Program Users Guide," Westinghouse, 1976.

(15) "IEEE Subcommittee Report on Excitation Systems," _Stability of Large Electric Power Systems_, IEEE Press, 1974.

(16) EPRI, "Coherency Based Equivalents for Transient Stability Studies," Report 904, Jan. 1975.

(17) Young, C. C., "The Synchronous Machine," _Modern Concepts in Power Systems Dynamics_, IEEE Press, 70M62-PWR, 1970.

Power Control System and Protection

CONTROL OF WIND TURBINE GENERATORS CONNECTED TO POWER SYSTEMS

H.H. Hwang and H.V. Mozeico

University of Hawaii

L.J. Gilbert

NASA-Lewis Research Center

I. INTRODUCTION

A. Historical Review

    Amidst the current crisis of dwindling organic fuel supplies, the prospect of power from the bountiful wind is an enticing one. Pioneering the generation of electrical energy by wind power was a profusion of designs and devices implemented in the 1920's and 1930's. Madaras' experiment, Ref. (1), with a rotating cylinder and Savonius' wing rotor, Ref. (2), exemplified some of the exotic designs which yielded to the more economical high-speed propeller type turbines. Examples of the latter are Darrieus' 2-bladed unit in France, Ref. (3), an operational Russian turbine, Ref. (4), driving a 100-kw induction generator, the briefly operational Smith-Putnam 1250-kw unit in the United States, Ref. (5), and more recently Hütter's 100-kw unit in Europe, Ref. (6).

    The control of a system which generates power from so unsteady an input as the wind presents a formidable problem. The speed varies from instant to instant due to gusts, and is further disturbed by the obstruction due to the supporting tower. In an effort to stabilize the effects of such an input upon the mechanical stresses of the turbine and the electromagnetic transients of the generator, attempts have been made to keep the rotor speed as nearly constant as possible. Perhaps the most promising method to control the rotor speed or the output of the system is the one employed by the Smith-Putnam machine, the Russian unit, and Hütter's machine: control the pitch of the blades relative to their plane of rotation. However, this method demands a sophisticated feedback control system. Unfortunately, even the choice of controlled variable to which the feedback system responds is at issue. For example, the blade pitch control of the Smith-Putnam and Russian machines responded to the rotor speed, while Hütter's mechanism was regulated by the output power.

B. ERDA-NASA Wind Energy Project

Current research in wind turbine technology is dominated by the ERDA-NASA wind energy program. ERDA-NASA has planned a graduated wind turbine project, with turbines ranging from 100-kw to 1500-kw, Ref. (7). The project currently has a 100-kw wind turbine at the NASA Plume Brook Station near Sandusky, Ohio, which became operational in 1975 (Fig. 1). This wind turbine, designated the Mod-0, is a horizontal axis, propeller type machine. A two bladed, 125 foot diameter rotor, mounted on a 100 foot truss tower, drives an alternator through a step-up gear box. The rotor is downwind of the tower and rotates at a constant speed of 40 rpm. The alternator is a 125 kva, 3 phase, 60 hz, 1800 rpm, 480 volt, Y-connected synchronous machine. Figure 2 shows details of the wind turbine generator system and assembly. Blade pitch control has been proposed, and implemented, as the control mechanism for the ERDA-NASA turbines. Figure 3 represents the blade pitch control system

Fig. 1   ERDA-NASA 100-kw Wind Turbine

for the Mod-0. This simplified representation includes transfer functions describing the blade torsional dynamics and transfer functions describing the pitch change mechanism. A speed feedback is employed for starting and synchronizing the generator, and output power feedback for system stability under load conditions.

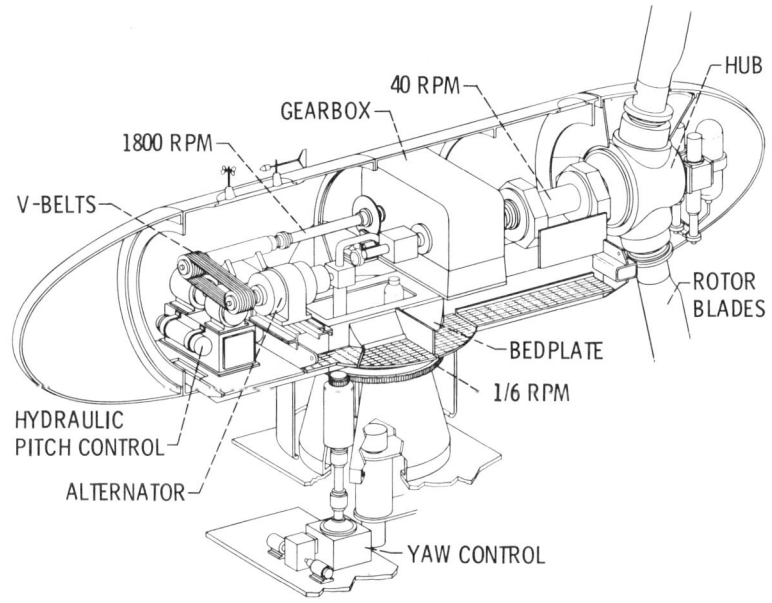

Fig. 2 100-kilowatt Wind Turbine Drive Train Assembly and Yaw System

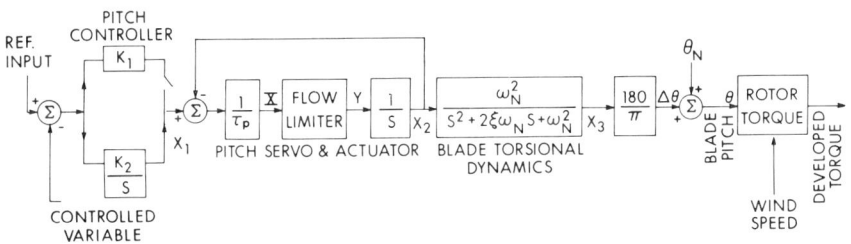

Fig. 3 Blade Pitch Control System of the ERDA-NASA 100-kw Wind Turbine

II. CONTROL STUDIES

As noted above, the implementation of a blade pitch control is a delicate problem, with only a few attempts at analysis of such systems to date. Hwang and Gilbert, Refs. (8,9), have studied the synchronization of the Mod-O wind turbine against an infinite bus. Johnson and Smith, Ref. (10), did a study of the effects of various system configurations under gusting conditions upon the power and current variations of an idealization of data for the Mod-0. However, further analysis may perhaps facilitate the design and successful utilization of these blade pitch control mechanisms.

This study presents an analytical representation for a wind turbine which employs blade pitch angle feedback control. A mathematical model will be formulated. With the functioning Mod-0 serving as a practical case study, results of a computer simulation of the model as applied to the problems of synchronization and dynamic stability will be presented. Figure 4 is a schematic of the system under study.

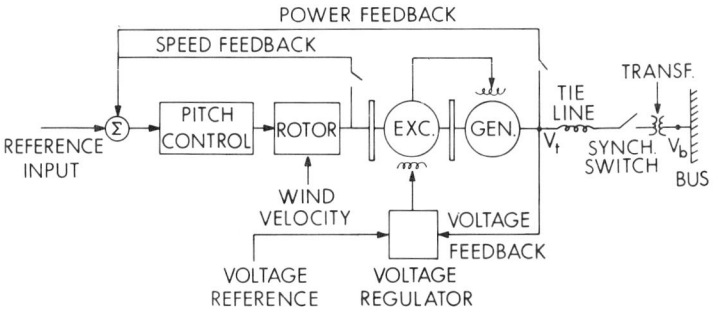

Fig. 4 A Schematic of the System Under Study

A. Speed Control

For the Mod-0, the rotor speed is the controlling variable during the starting up and synchronization operations. Here the feedback loop uses the ideal integral plus a proportional compensator. Speed control will be investigated here during the synchronization process. Since synchronization can be severely impeded by wind gusts, a gust is superimposed onto the mean wind speed during this simulation.

B. Power Control

To stabilize the output under load conditions, the Mod-0 feedsback the power output of the alternator to the blade pitch control system. The pitch controller, as indicated in Fig. 3, consists only of an integrator, without the proportional component ($K_1 = 0$). Practical experience with the Mod-O has revealed, Ref. (11), that a significant, periodic perturbation occurs during this operation caused by the obstruction of the wind by the supporting tower of the turbine. The rotor blades, which are downwind of the tower, enter into this 'tower shadow' of reduced wind velocity during each revolution (Fig. 5), Ref. (12). The simulation of the Mod-O's power control includes this tower shadow effect.

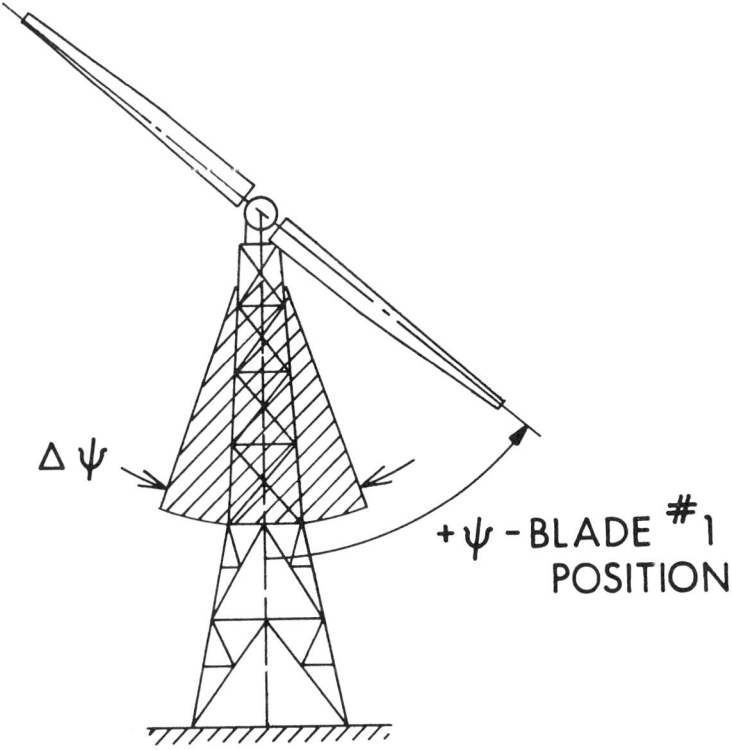

Fig. 5 The Tower Shadow

III. SYSTEM EQUATIONS

A. Synchronous Machine Equations

Park's equations, Ref. (13), with appropriate modifications, Refs. (9,14), are used. It is reasonable to assume that the transformer e.m.f.s. $(\frac{d\lambda_d}{dt}, \frac{d\lambda_q}{dt})$ are negligible as compared to the speed voltages $(\omega_o\lambda_d, \omega_o\lambda_q)$. Also, the synchronous generator frequency may be assumed to be the same as the system frequency, ω_o, as long as synchronism is retained.

$$\frac{de'_q}{dt} = (V_f - e_{q1})/T'_{do} \tag{1}$$

$$\frac{de''_q}{dt} = -\frac{e_{q2}}{T''_{do}} \frac{x'_d - x''_d}{x_d - x''_d} \tag{2}$$

$$\frac{de''_d}{dt} = -e_d/T''_{qo} \tag{3}$$

$$i_d = \frac{(R_a e''_d + x''_q e''_q) - (R_a V_d + x''_q V_q)}{R_a^2 + x''_d x''_q} \tag{4}$$

$$i_q = \frac{(R_a e''_q - x''_d e''_d) - (R_a V_q - x''_d V_d)}{R_a^2 + x''_d x''_q} \tag{5}$$

$$\lambda_d = e''_q - i_d x''_d \tag{6}$$

$$\lambda_q = -e''_d - i_q x''_q \tag{7}$$

$$e_d = \frac{x_q}{x''_q} e''_d + \frac{(x_q - x''_q)}{x''_q} \lambda_q \tag{8}$$

$$e_{q1} = \frac{x_d - x''_d}{x'_d - x''_d} e'_q - \frac{x_d - x'_d}{x'_d - x''_d} e''_q \tag{9}$$

$$e_{q2} = -\frac{x_d - x''_d}{x'_d - x''_d} e'_q + \frac{x'_d}{x''_d} \frac{x_d - x''_d}{x'_d - x''_d} e''_q - \frac{x_d - x''_d}{x''_d} \lambda_d \tag{10}$$

The armature current is given by

$$i_a = \sqrt{i_d^2 + i_q^2} \tag{11}$$

The electromagnetic torque is:

$$T_e = [e''_q i_q + e''_d i_d - i_d i_q (x''_d - x''_q)] \tag{12}$$

Equations (1-12) are in per unit based upon the machine rating. In these equations, the total tie line and transformer reactance, X_L, are included in the machine impedances. Therefore, voltages V_d and V_q in Eqs. (4,5) are the d-axis and q-axis components of the bus voltage unless stated otherwise.

B. Equations of Motion

The electromagnetic torque (referred to the rotor shaft) in lb-ft is

$$T_L = \frac{NP}{2} \frac{1}{\omega_o} \text{(kva base)} (738) T_e \qquad (13)$$

The equations of motion, or swing equations, are

$$\frac{d\delta_m}{dt} = \Omega - \Omega_N - \Omega_e \qquad (14)$$

$$\frac{d\Omega}{dt} = \frac{(T-T_L) - B\Omega}{J} \qquad (15)$$

where δ_m is in mechanical radians and T is in lb-ft.

The torque angle of the alternator in electrical degrees is given by

$$\delta = \frac{180}{\pi} \frac{NP}{2} \delta_m \qquad (16)$$

C. Equations of the Blade Pitch Control System

1. Speed Control

For speed control, the controlled variable and reference input are the instantaneous rotor shaft speed, Ω, and the corresponding synchronous speed, Ω_N, respectively. The following equations are formulated according to Fig. 3:

$$\frac{dX_1}{dt} = -K_2 \Omega_e \qquad (17)$$

$$\frac{dX_2}{dt} = Y \qquad (18)$$

$$\frac{dX_3}{dt} = X_4 \qquad (19)$$

$$\frac{dX_4}{dt} = \omega_N^2 X_2 - \omega_N^2 X_3 - 2\xi\omega_N X_4 \qquad (20)$$

The input and output variables of the flow limiter are

$$X = \frac{1}{\tau_p} (X_1 - K_1 \Omega_e - X_2) \tag{21}$$

$$Y = \text{limit}(-\frac{\pi}{22.5}, \frac{\pi}{22.5}, X) \tag{22}$$

The change in pitch angle is:

$$\Delta\theta = \frac{180}{\pi} X_3 \tag{23}$$

The pitch angle is

$$\theta = \theta_N + \Delta\theta \tag{24}$$

The corresponding torque developed by the rotor in lb-ft is:

$$T = f(V_w, \Omega_r, \theta) \tag{25}$$

Equation (25) is a nonlinear function of wind speed in mph, rotor speed in rpm, and pitch angle in mechanical degrees, Ref. (15). The turbine torque-pitch angle curves are given in Fig. 6. The blade pitch angle physically can vary only from -90 degrees to zero degrees so that the slope of the torque angle curves is always positive.

2. Power Control

The controlled variable and the reference input are the instantaneous power output of the generator, p_o, and the rated power output p_N, in kilowatts, respectively.

Equation (17) in this case becomes

$$\frac{dX_1}{dt} = K_2'(P_N - P_o) \tag{26}$$

The value of K_1 is equal to zero in Eq. (21). Other equations remain the same.

The power output of the generator in kilwatts is

$$p_o = (\text{kva base})(V_d i_d + V_q i_q) \tag{27}$$

D. Equations of Generator Excitation Control System

A static excitation regulator is used to hold the terminal voltage, the output power factor, and the armature current

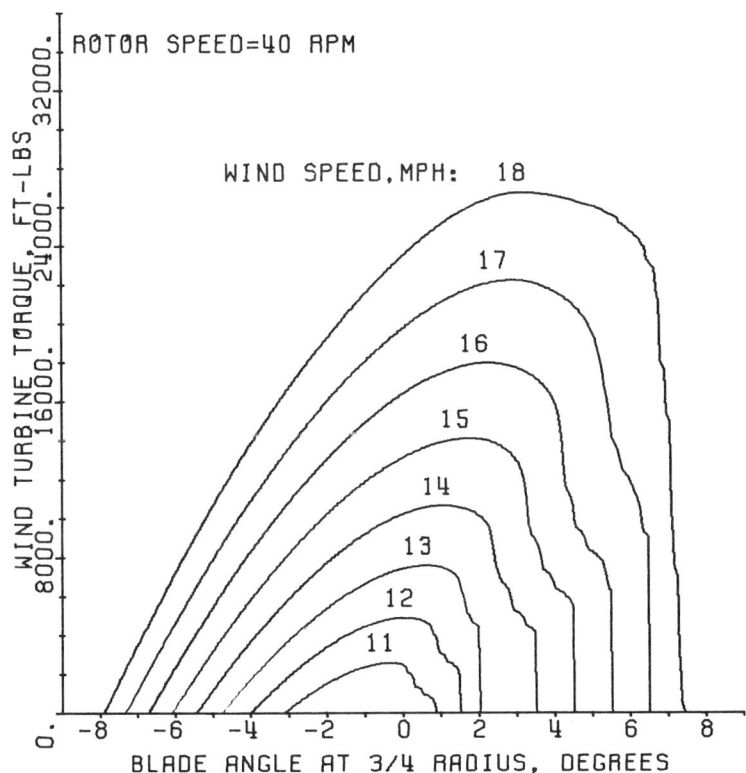

Fig. 6 The Turbine Torque-Pitch Angle Curves

of the generator within satisfactory limits. The excitation control system is approximated by the following equations:

$$\frac{d\Delta V_f}{dt} = z \tag{28}$$

$$\frac{dz}{dt} = \frac{\mu_e}{\tau_e} \Delta V_t - \frac{z}{\tau_e} \tag{29}$$

$$V_f = V_{fN} + \Delta V_f \tag{30}$$

$$\Delta V_t = (V_N - V_t) - K_3(0.8\ i_a - \frac{P_o}{V_t(\text{kva base})}) \tag{31}$$

$$V_{td} = V_d - X_L i_q \tag{32}$$

$$V_{tq} = V_q + X_L i_d \tag{33}$$

$$V_t = \sqrt{V_{td}^2 + V_{tq}^2} \tag{34}$$

The first term in Eq. (31) corrects for the deviation of the generator terminal voltage, V_t, from its reference value, V_N. The second term corrects for power factor deviation. At rated output, the practical power factor is equal to 0.8, lagging current. V_{td} and V_{tq} in Eqs. (32,33) are the d-axis and q-axis components of V_t.

E. Numerical Method

The system equations formulated in the preceding sections are nonlinear and interrelated. They are solved numerically by the fourth order Adams-Bashforth (predictor) and Adams-Moulton (corrector) method, for which a fixed stepsize of h=0.005 is used.

IV. NUMERICAL EXAMPLES

A. System Constants for the Mod-0

1. Synchronous Machine and Electrical Constants

   R_a=0.018, x_d=2.21, x_q=1.064, x'_d=0.165 p.u., x''_d=0.128, x''_q=0.193, V_b=1.000, x_L=0.02 p.u., T'_{do}=1.94212, T''_{do}=0.01096, T''_{qo}=0.06230 sec.

2. Mechanical Constants

   N=45, J=10^5 lb-ft-sec^2, B=959 lb-ft-sec.

3. Control System Constants

   K_1=5 sec, K_2=0.107 per sec, K'_2=6 x 10^{-4} per kw-sec, ω_N=100 radians per sec, $\tau_p = \dfrac{1}{2 \times 2.7 \times \pi}$ sec, ξ=0.02

4. Excitation Control System Constants

   μ_e=30.0000, T_e=0.0500 sec, K_3=0.2000

B. Example of Speed Control

The problem of synchronizing the Mod-O generator against an infinite bus under gusting wind conditions is simulated here as an example.

1. Equation of Discrete Wind Gust

The following equations for estimating the discrete longitudinal gust were recommended by G.H. Fichtl at the NASA-George C. Marshall Space Flight Center:

$$V(t,z) = \frac{A(\tau,z)}{2} (1 - \cos \frac{2\pi t}{\tau}), \quad 0 \leq t \leq \tau \quad (35)$$

where A is the gust amplitude and τ is the gust period. The gust amplitude is a function of the gust period τ and altitude z, namely

$$A = 3 \frac{u_r}{\ln(z_r/z_o)} \{1-\exp[-U(z_a)\tau/1.48\, z_a]\}^{1/2} \quad (36)$$

in which $U(z_a)$ is the mean wind at z_a above natural grade, U_r is the mean wind at a reference level z_r and z_o is the surface roughness length. For the Mod-O, $z_r=z_a=100$ ft, $z_o=0.656$ ft.

2. Digital Simulation

Two initial speed errors, 7.5 and -7.5 on an 1800 rpm basis, are used. They are the extreme limits of the synchronizing zone of the existing synchronizer. Procedures for bringing the machine to these speeds are outlined in a previous publication, Ref. (16).

A mean wind speed of 15 mph is assumed. The gust period is equal to about 0.5 seconds. Using Eqs. (35,36), the amplitude of the gust is estimated to be 15% of the mean wind speed.

The change in voltages due to closing the synchronizing breaker contacts are

$$\Delta V_d = \sin \delta \quad (37)$$

$$\Delta V_q = \cos \delta - 1 \quad (38)$$

The net effect of closing the breaker contacts is simulated by replacing V_d by ΔV_d in Eq. (4) and V_q by ΔV_q in Eq. (5), and setting the field voltage V_f to zero in Eq. (1). The initial value of field current is included in the numerical computations of electromagnetic torque given by Eq. (12). As the reactance X_L between the generator terminal and the bus is very small compared to other constants, the terminal voltage, V_t, practically remains constant. Therefore, the excitation control system has very little effect on synchronization and is omitted in the simulation.

3. Results of Computer Studies

The following results are observed:

Figure 7(a): The initial rotor speeds referred to the low speed shaft are 40.1667 rpm for case (a) and 39.8333 rpm for case (b). The rotor speed reaches its synchronous speed (40 rpm) within an error less than 0.1 percent at a time equal to about 0.70 seconds for case (a), and 0.9 seconds for case (b), immediately following the synchronization switching. For both cases, synchronous speed is reached within the first two seconds after closing the synchronizing switch.

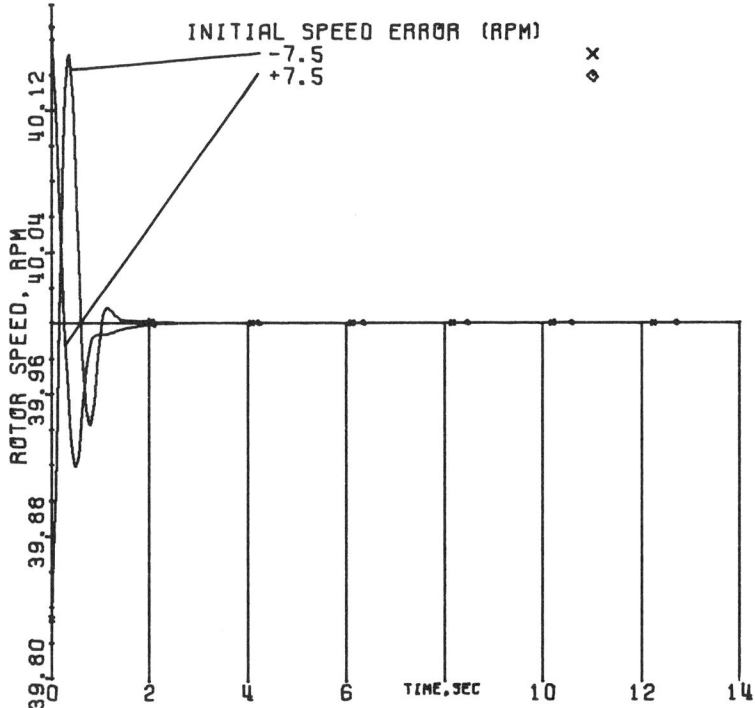

Fig. 7(a)   Wind Turbine Rotor Speed

Figure 7(b): The plot of generator power or torque angle vs. time (or swing curve) is the best indication of the system stability under transient conditions. The maximum and minimum power angles, for case (a), are about 17 electrical degrees (at t=0.26 sec) and 0.3 electrical degrees (at t=4.66 sec). For case (b), the corresponding values are 5.8 (at t=0.62 sec) and -17 (at t=0.16 sec). After 2 simulated seconds the absolute value of the power angle for either case is less than 0.5 electrical degrees. The small variation and fast convergence of the power angle assures that a smooth synchronization can be achieved under the assumed conditions.

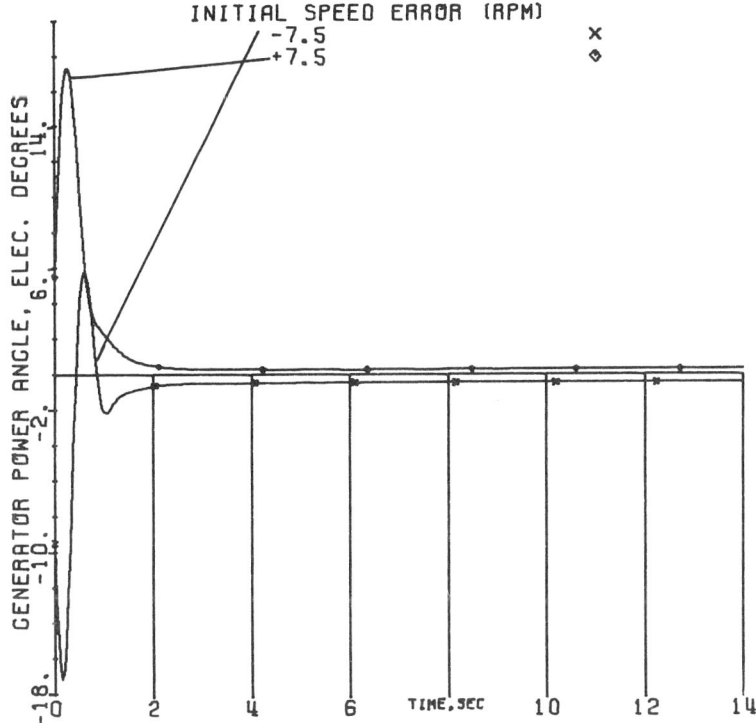

Fig. 7(b)   Generator Power Angle

Figure 7(c): The curves given in this figure show the variation of the wind turbine blade pitch angle. The wind turbine torque and rotor speed are controlled by varying the blade pitch angle. Experiments have recently been performed on the Mod-0 generator under similar initial rotor speed and wind conditions. The computer results given here matched experimental results very closely.

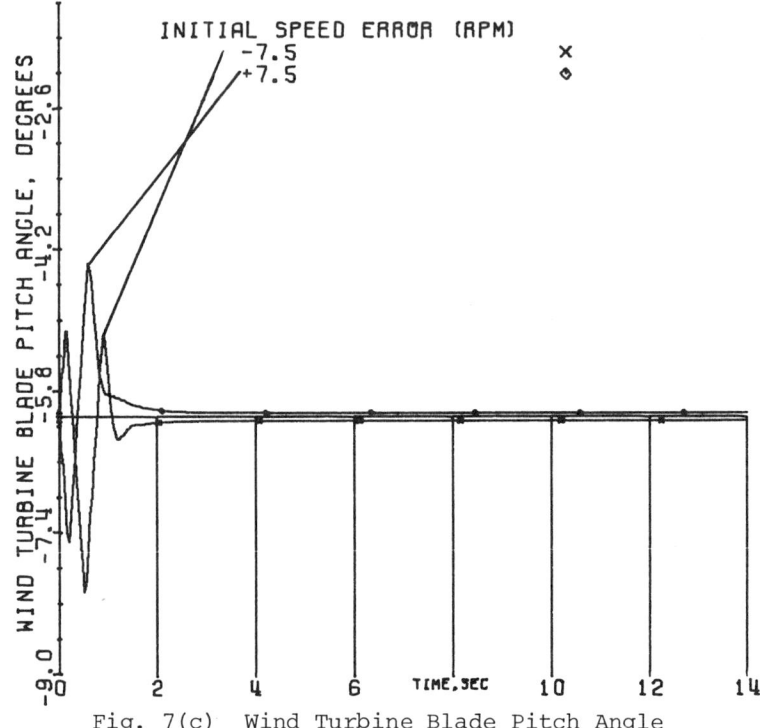

Fig. 7(c)   Wind Turbine Blade Pitch Angle

Owing to space limitation, curves of armature current, electromagnetic torque, and wind turbine torque are not given here. However, as given by computer results, the corresponding transients are reasonably small.

C. Example of Power Control

System stability at rated output (100 kw) is simulated here as an example.

1. Tower Shadow Effect

Data from the NASA-Lewis Research Center reveals that the Mod-0 sustains approximately a 28% decrease in torque as a blade passes through the tower shadow. The shadow's arc is estimated to be 30 degrees. Using Eq. (25) and holding both the rotor speed and the blade pitch angle constant, the 28% decrease in torque is equivalent to a 10% decrease in wind velocity in the shadow.

A cosine function models the tower shadow effect, and simplifies the 2 physical blades rotating through a full revolution to 1 blade rotating through a half revolution.

2. Digital Simulation

Initially, the wind generator is supplying rated power (100 kw) to the bus at 0.8 power factor, lagging current, and rated frequency. An 18 mph mean wind speed with a 10% tower shadow reduction is used in this simulation. It is further assumed that the tower shadow cuts in at t=0.

Field excitation control system is implemented and included in the simulation. The d-axis and q-axis components of the bus voltage, in this case, are:

$$V_d = \sin \delta \qquad (39)$$

$$V_q = \cos \delta \qquad (40)$$

Values of V_d and V_q are used in Eqs. (4,5).

3. Results of Computer Studies

Figure 8(a): The plot of the effective wind speed at the turbine's blades, for one revolution. The vertical axis is the wind speed in mph. The decreased wind velocity occurs when one blade is in the tower shadow.

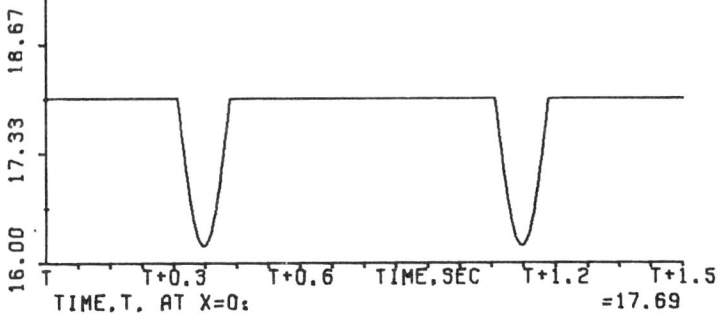

Fig. 8(a) Wind Velocity, mph vs. time for one revolution

Figure 8(b): The plot of the turbine torque for one revolution of the blades. The reduction caused by the tower shadow is about 28 percent. The vertical axis is the wind turbine torque in ft-lbs.

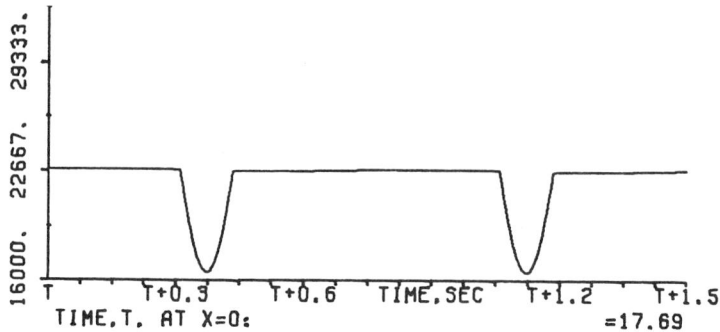

Fig. 8(b) Turbine Torque, ft-lbs vs. time for one revolution

Figure 8(c): The plot of the alternator output. The effect of abruptly superimposing the tower shadow on the initially stable system is clearly visible. Recall that the power output is the controlling variable.

The minimum value of the alternator output, about 85 kw, occurs about 0.12 seconds after the first blade has left the tower shadow. The output then oscillates, with the maximum value for each cycle occurring just prior to or just as entering a 'new' shadow. After the first cycle, the average output gradually increases an additional 2 kw until time 6 seconds. The output then stabilizes between 112 kw and 88 kw. The variation is only 12 percent above or below the mean value.

Figure 8(d): The plot of the turbine blade pitch angle. Recall changes in blade pitch angle occur in response to the alternator output. The blade angle is constant through the first 0.06 seconds, then begins to increase in response to the decreasing power output. The angle continues to increase until about 0.06 seconds after the output (now increasing) climbs beyond its reference value. The angle then decreases until 0.06 seconds after the output (now decreasing) falls below its reference value. The 0.06 second lag in blade pitch response to the matching of the alternator output with its reference value continues throughout the simulation.

Figure 8(e): The plot of wind turbine rotor speed. The rotor speed, initially 40 rpm, has a minimum value of 39.95 rpm at time 0.88 seconds. Thereafter, the rotor speed remains within about .1 percent of 40 rpm.

Figure 8(f): The plot of armature current. The initial value of the armature current is 0.9884 p.u. Its maximum and

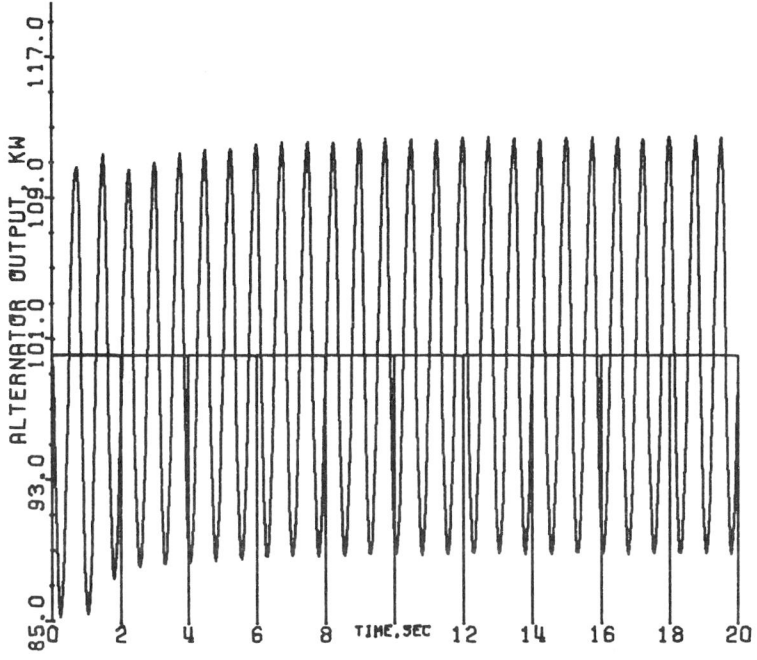

Fig. 8(c)   Alternator Output

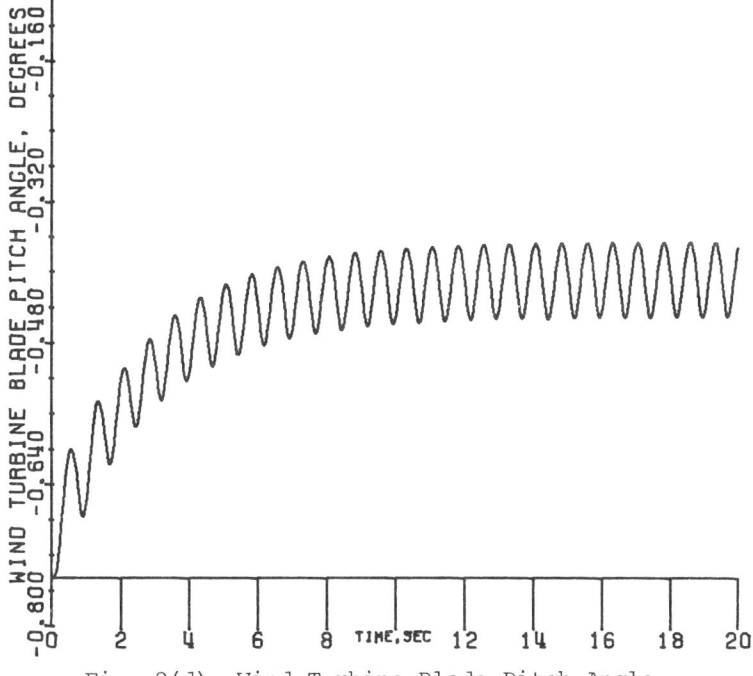

Fig. 8(d)   Wind Turbine Blade Pitch Angle

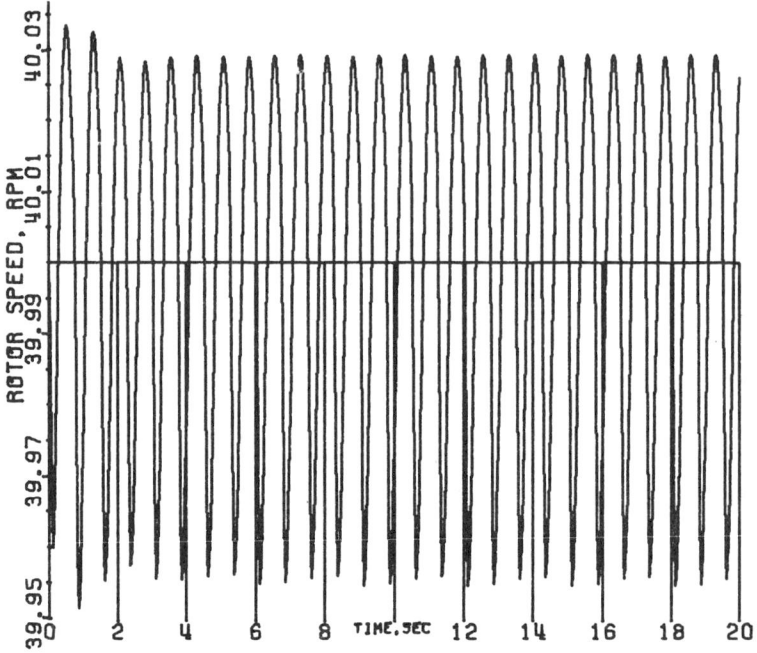

Fig. 8(e) Wind Turbine Rotor Speed

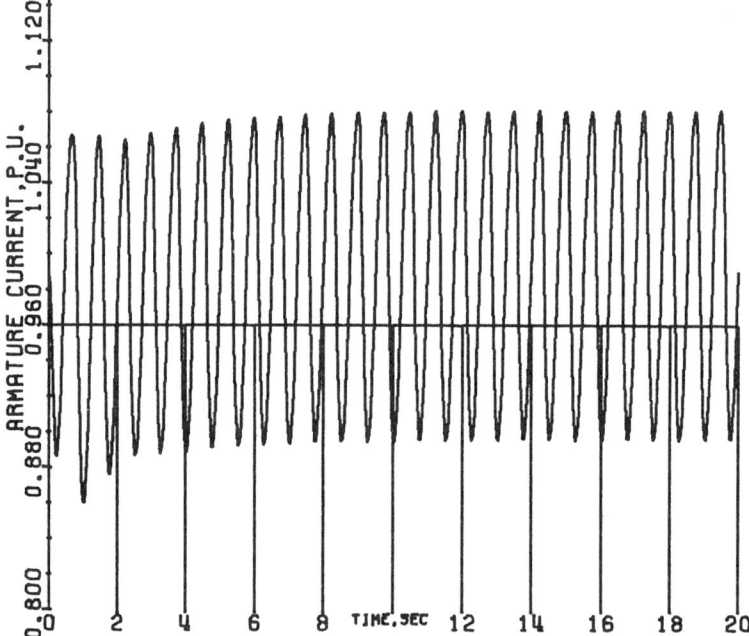

Fig. 8(f) Armature Current

minimum values are 1.0814 (at t=18.74 sec) and 0.8589 (at t=1.02 sec), respectively.

Figure 8(g): The plot of the generator power angle. The initial value of the torque angle is 27.53 degrees. Maximum and minimum values are 30.04 (at t=1.54 sec) and 24.16 (at t=0.30 sec), respectively.

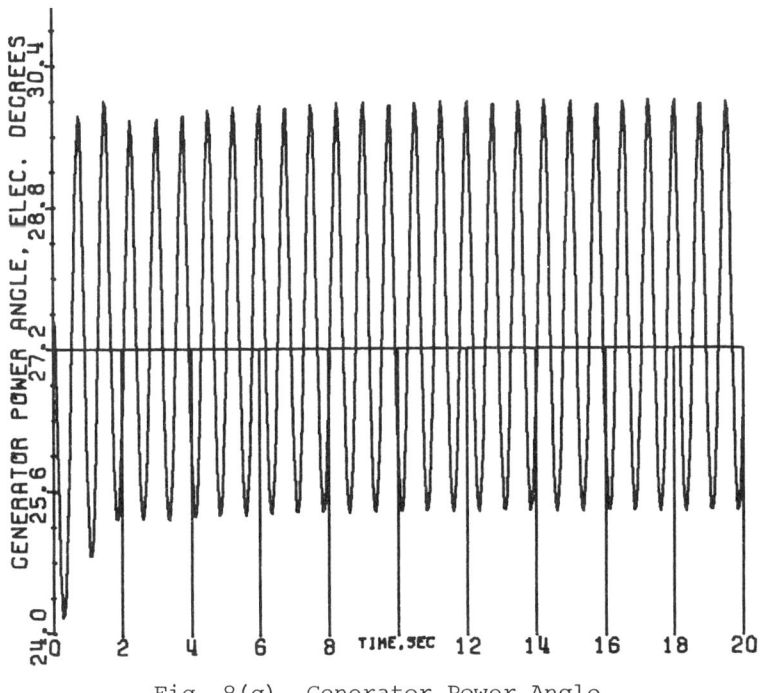

Fig. 8(g) Generator Power Angle

The computer results reveal that the output power factor, not shown here, varies between 0.824 and 0.778 once the system has stabilized.

V. CONCLUSIONS

A unique simulation model has been developed according to the Mod-0 wind turbine. This model can be used for simulating both speed and power control.

Both experimental and computer simulation results reveal that the speed and power output of a wind turbine can be satisfactorily controlled within reasonable limits by employing the existing blade pitch control system under the conditions given in sections IV-A and IV-B.

For power control, an additional excitation control is needed so that the terminal voltage, output power factor, and armature current can be held within narrow limits. As a consequence, the variation of torque angle is limited even if speed control is not implemented simultaneous to the power control.

VI. NOMENCLATURE

The symbols used correspond insofar as possible to standard definitions. Other symbols used are

λ_f, λ_{11q}	flux linkage of field and damper circuit in j axis
M_{af}, M_{alj}	maximum mutual inductance between armature circuit and field (or damper circuit in j axis)
L_{ff}, L_{11j}	self inductance of field circuit and damper circuit in j axis, respectively
ω_o	base angular velocity
P	number of poles of the synchronous machine
N	gear ratio between high and low speed shaft
V_{fN}, V_f	reference or instantaneous field voltage referred to the stator.
B	damping coefficient of the complete rotating system in lb-ft-sec
J	inertia of the complete rotating system (referred to low speed shaft) in lb-ft-sec^2
Ω	angular speed of wind turbine rotor in radians per second
Ω_N	reference speed (referred to wind turbine rotor shaft) in radians per second
K_1	controller proportional gain in seconds
K_2, K_2'	controller integral gain per second or per kw-sec.
θ	pitch angle of blade in mechanical radians
θ_N	reference pitch angle of blade in mechanical radians
ξ	damping ratio
τ_p	control hydraulic actuator time constant in seconds
Ω_e	rotor rotational speed error in radians per second
ω_N	undamped natural angular frequency in radians per second

$$e_{q1} = \omega_o M_{af} i_f, \quad e_{q2} = \omega_o M_{ald} i_{11d}, \quad e_d = -\omega_o M_{alq} i_{11q}$$

$$e_q'' = \omega_o \frac{M_{ald}}{L_{11d}} \lambda_{11d}, \quad e_q' = \omega_o \frac{M_{af}}{L_{ff}} \lambda_f, \quad e_d'' = -\omega_o \frac{M_{alq}}{L_{11q}} \lambda_{11q}$$

VII. ACKNOWLEDGMENTS

We are grateful to Dr. W.A. Lewis for his valuable suggestions. Thanks to the Center for Engineering Research of the University of Hawaii for typing this manuscript. This work is supported by the NASA-Lewis Research Center and the Hawaii Natural Energy Institute.

VIII. REFERENCES

(1) The Electrician, (Nov. 24, 1933).
(2) Savonius, S.J., Mech. Eng., (May, 1931).
(3) Darrieus, La Nature, (December 15, 1929).
(4) Sectorov, W.R., Elektrichestvo 2, (1933).
(5) Putnam, P.C., "Power from the Wind," Van Nostrand Reinhold, New York, 1948.
(6) Hütter, V., NASA Tech. Trans. F-15-068, (1973).
(7) Thomas, R.L., NASA TMX-71890, (1976).
(8) Hwang, H.H., and Gilbert, L.J., in "1977 Control of Power Systems Conference and Exposition Conference Record," p. 26, IEEE, New York, 1977.
(9) _____, in "IEEE PICA-77 Conference Proceedings," p. 326, IEEE, New York, 1977.
(10) Johnson, C.C., and Smith, R.T., IEEE Trans. on Aerospace and Electronic Systems, Vol. AES 12, No. 4, 483, (1976).
(11) Savino, J.M., and Wagner, L.H., NASA TMX-73548, (1976).
(12) Glasgow, J.C., and Linscott, B.S., NASA TMX-71601, (1976).
(13) Park, R.H., AIEE Trans. 48, 716, (1929).
(14) Oliver, D.W., IEEE Trans. on Power Apparatus and Systems 87, 1669, (1968).
(15) Wilson, R.E., and Lissaman, P.B.S., "Applied Aerodynamics of Wind Power Machines," Oregon State University, 1974.
(16) Gilbert, L.J., NASA TMX-71864, (1976).

Power Control System and Protection

OPERATION AND CONTROL OF WIND-ELECTRIC SYSTEMS

R. Ramakumar

Oklahoma State University

I. INTRODUCTION

Realization of the limitations of fossil fuels has spurred world-wide interest in the utilization of renewable energy sources [1-4]. All renewable energy resources trace back to the sun and wind energy is one manifestation of solar energy that has attracted the attention of human beings for thousands of years [5].

The primary application envisaged for wind is in the generation of electrical energy [6-8]. The ease with which aeroturbines transform energy in moving air to rotary mechanical energy suggests the use of electrical devices to convert wind energy to electrical form for a variety of end uses.

Large-scale utilization of wind energy is expected to be in the fuel conserving mode [9,10]. Wind-electric systems convert energy in moving air to constant frequency ac form to be fed into existing utility lines to augment the total energy generated and save fuel. This procedure earns 'energy credit' for wind-electric systems. Several wind driven generator units, distributed over a large geographical region, may yield some firm capacity depending on the diversity in wind regimes at different locations. Under these circumstances, wind-electric systems earn both 'energy credit' and some 'capacity credit'.

Wind-electric systems generate power only when the wind blows. Therefore, they can be classified as 'intermittent base load' devices. Some kind of back-up capacity is required to provide adequate reliability and meet the system demand. This back-up capacity could be provided by other generators tied to the utility grid or by a separate gas turbine generator system operating in parallel with the wind-electric units.

This paper discusses the operational and control aspects of wind-electric systems. A brief review of significant past work, future trends and research needs are included.

II. OPERATION FUNDAMENTALS

A. General

Power density in moving air is proportional to the cube of wind speed as given below.

$$P_w = K_w V^3 \qquad (1)$$

The constant K_w depends on the units used and is equal to 0.6386 when area is measured in m^2, wind speed in m/s and power density in W/m^2. However, only a fraction of this can be converted to mechanical shaft power by an aeroturbine. This fraction is known as the coefficient of performance (or power coefficient) C_p. Figure 1 illustrates the dependence of this power coefficient on tip-speed to wind-speed ratio λ for several typical aeroturbines.

For a swept area of A, the electrical power output P_e is given by:

$$P_e = \eta_g \eta_m A C_p K_w V^3 \qquad (2)$$

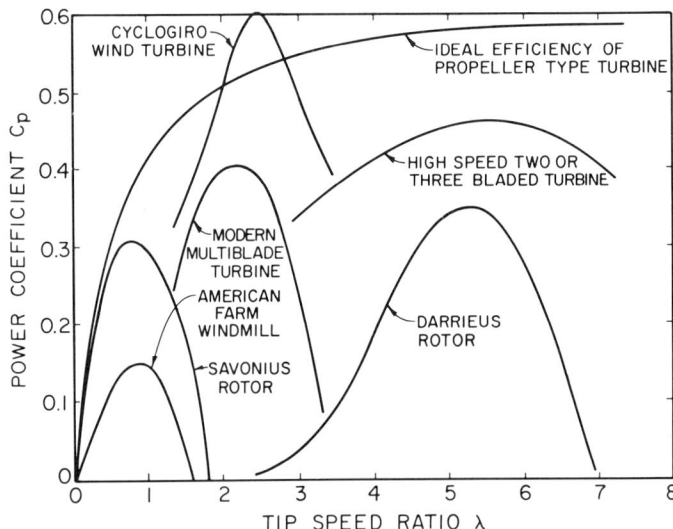

Figure 1. Illustrating C_p vs λ Characteristis of Aeroturbines.

where η_g and η_m are the efficiencies of electrical generator and mechanical gearing, respectively.

With constant speed operation, changes in wind speed change λ and consequently C_p. However, reasonably high values of C_p are achieved throughout the operating range by suitably changing the pitch angle of blades. Once the generator output reaches its rated value, it is maintained constant at that value for further increases in wind speeds by suitable pitch control. In addition, for extreme wind speeds, pitch control furls the blades to avoid damage. Thus, pitch control becomes an essential and important ingredient of constant-speed wind-electric systems.

In variable speed operation, the tip-speed ratio is maintained at the desired optimum value (that maximizes C_p) throughout the operating range of wind speeds by appropriately adjusting the output of the system. With a constant C_p, the following relationships must be valid:

$$\omega \propto V \qquad (3)$$

$$T \propto V^2 \qquad (4)$$

where ω and T are the aeroturbine angular speed and shaft torque, respectively. Even in this case, a simple pitch control mechanism is necessary to furl the blades at extreme windspeeds.

B. Constant vs Variable Speed Operation

Variable-speed wind-electric systems have the potential to extract a larger fraction of the energy in the wind. This can be seen by referring to Figure 2 which shows typical wind speed-duration and power-duration curves [11]. The wind-electric system starts delivering power at a wind speed called the cut-in speed V_C and the plant is shut down for wind speeds above a safe value called the furling speed V_F. For intermediate values of wind speeds, output power is determined by the coefficient of performance of the aeroturbine and the efficiencies of the mechanical interface and the electrical generator.

At low wind speeds, constant-speed systems operate at a higher tip-speed ratio. Such an operation, in spite of appropriate changes in pitch angle of blades, results in C_p values that are less than optimum. Variable-speed systems operate at a constant tip-speed ratio and consequently can maintain a high C_p even at low wind speeds. This accounts for the higher outputs at low wind speeds.

At high wind speeds, once the generator output reaches its rated value, it is maintained at that value for any additional

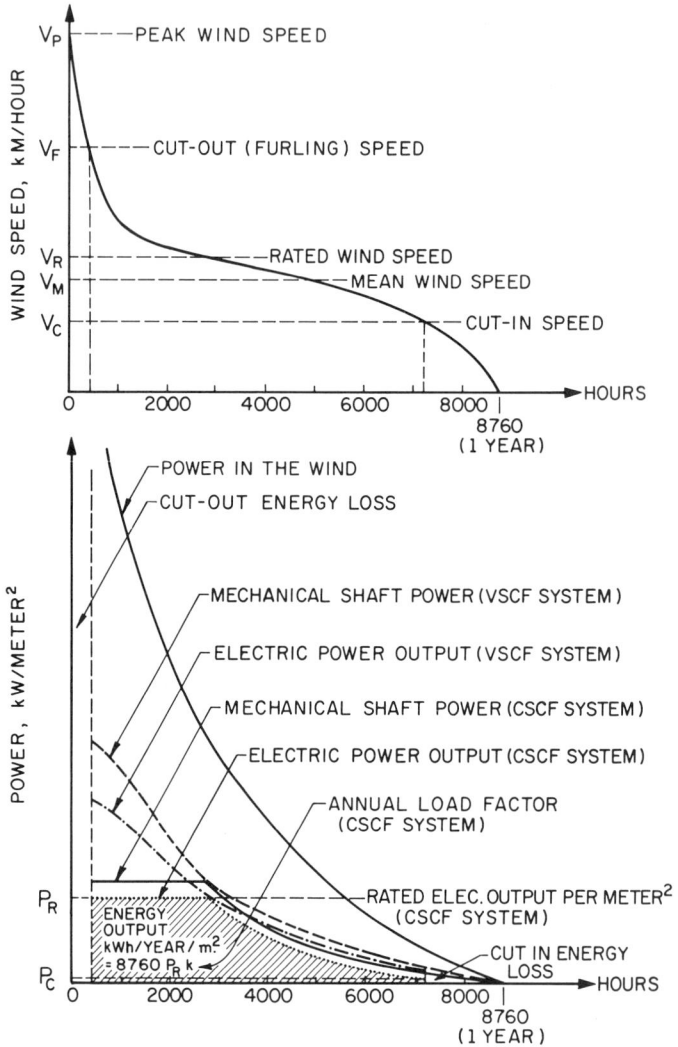

Figure 2. Typical Wind Speed and Power Duration Curves.

increases in wind speeds up to the furling speed in the case of constant speed systems. This process spills valuable energy in wind at high wind speeds. Spilling is tolerated to avoid overloading the electrical generator and to mitigate the electrical stability problems under constantly varying input conditions. The latter is particularly aggravated by bi-directional power flow between electrical grid and generator.

However, many variable-speed constant-frequency generation schemes (discussed later) allow power flow only in one direction - from the generator to the grid. This characteristic may allow operation of variable-speed wind-electric systems at power outputs above rated value at high wind speeds as illustrated in Figure 2. The nature and extent of this increase will depend on the particular generation scheme employed and is not yet fully understood.

The discussion in the above two paragraphs leads one to conclude that variable-speed operation of wind-electric systems results in higher power outputs for both low wind speeds and high wind speeds. This leads to higher annual energy yields per rated installed kW. The percentage gain is highly dependent on the wind regime and the nature of the generation scheme employed. Both horizontal axis and vertical axis [12] wind turbines will exhibit this gain under variable-speed operation.

C. Basic Components

Figure 3 illustrates the basic components of a wind-electric conversion system. Aeroturbines convert energy in moving air to rotational mechanical energy. In general, they require

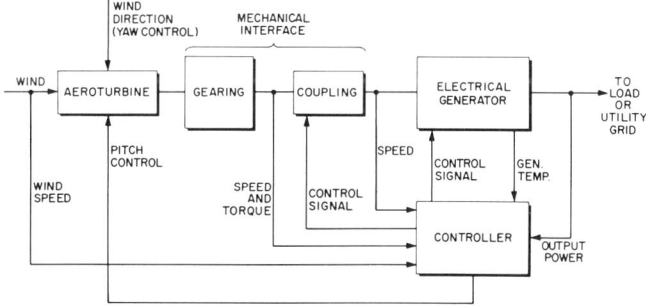

Figure 3. Basic Components of a Wind-Electric System.

pitch control and yaw control (only in the case of horizontal or wind-axis machines) for proper operation. A mechanical interface consisting of a step-up gear and a suitable coupling transmits the rotary mechanical energy to an electrical generator. The output of this generator is connected to the load or power grid as the application warrants.

The purpose of the controller is to sense wind speed, wind direction, shaft speeds and torques at one or more points, output power and generator temperature as necessary and generate appropriate control signals for matching the electrical output to the wind energy input and protect the system from extreme conditions brought upon by strong winds, electrical faults, and the like.

The coupling interfacing the high-speed shaft of the step-up gear box and the generator shaft plays a significant role in determining the behavior of a wind-electric system [13]. Several types of couplings are available for use. They are:

a. Rigid shaft coupling; angular strain of the shaft is negligibly small.

b. Stiff shaft coupling

$$T = K (\theta_1 - \theta_2) \tag{5}$$

where T is the torque transmitted, θ_1 and θ_2 the angular positions at the two ends of the shaft and K is the torsion constant.

c. Compliant shaft coupling

$$T = K (\theta_1 - \theta_2) \tag{6}$$

the value of K in this case is anywhere from 3 to 10% of the K in case 'b'.

d. Compliant shaft with damping

$$T = K (\theta_1 - \theta_2) + K_D \omega \tag{7}$$

where K_D is the damping constant and $\omega = (\omega_1 - \omega_2)$ where ω_1 and ω_2 are the angular speeds at the two ends of the coupling.

e. Rate couplings

(i) Eddy current coupling

Torque-slip characteristic of an eddy current coupling can be expressed in the following normalized form [14].

$$\frac{T_2}{T_r} = \frac{(2 + \sqrt{2})}{\sqrt{\frac{\omega}{\omega_m}} + \sqrt{\frac{\omega_m}{\omega}} + \sqrt{2}} \frac{T_m}{T_r} \qquad (8)$$

where T_2 = torque transmitted

T_r = rated torque (at slip ω_r)

T_m = maximum torque

ω = operating slip

ω_m = slip at maximum torque

Figure 4 shows a typical family of torque-slip characteristics for the eddy current coupling.

(ii) Fluid drive.

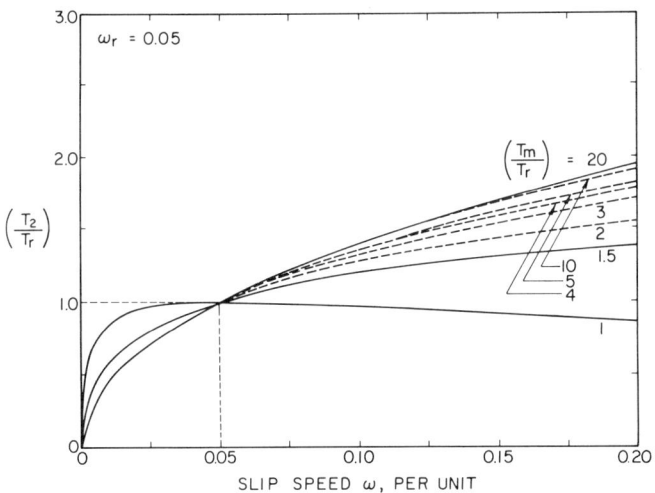

Figure 4. Torque-Slip Characteristics of Eddy-Current Couplings.

Figure 5 shows the torque-slip characteristic of a typical fluid drive coupling. These couplings limit the maximum torque transmitted to a predesigned value.

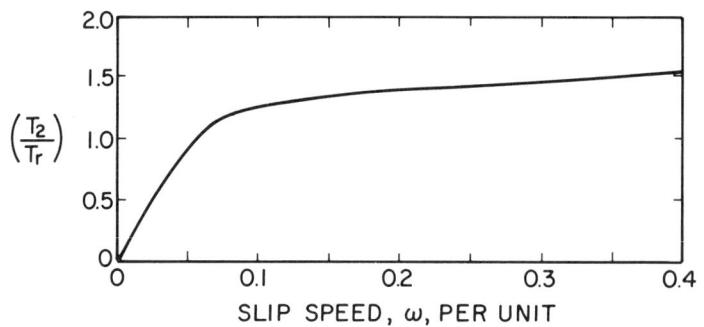

Figure 5. Torque-Slip Characteristic of a Typical Fluid Drive Coupling.

III. GENERATION SCHEMES

A. Classification

Generation schemes for wind-electric conversion are classified according to the type of output (constant frequency ac, variable frequency ac or dc) and mode of operation (variable speed, constant speed or nearly constant speed). Several combinations of speeds and outputs have been proposed for wind energy utilization, depending on the end use of the electrical energy generated. Each combination has its unique advantages and control problems.

The following schemes are being given serious consideration at present for large-scale wind-electric conversion.
    a. Constant speed shaft and constant frequency output.
        (i) Synchronous generators
    b. Nearly constant speed shaft and constant frequency output.

(i) Induction generators
c. Variable speed shaft and constant frequency output.
   (i) Field modulated generator system
   (ii) AC-DC-AC link.
   (iii) Double output induction generators

Since induction generators operate with a very small slip (2 to 5%), it is customary to group 'a' and 'b' and call them constant-speed constant-frequency systems.

The technology of synchronous and induction generators is mature and well known, especially to workers in the power area. Even the technology of AC-DC-AC link (see Figure 6) is well established with the introduction of high-voltage dc transmission systems in the utility grids of the world. Therefore, only field modulated generator system and double output induction generator will be discussed further in this section.

B. Field Modulated Generator System

A field modulated generator system (FMGS) employs a three-phase high-frequency alternator to generate single-phase power at the required low frequency [15-17]. The rotating field coil of the alternator is supplied with alternating current at

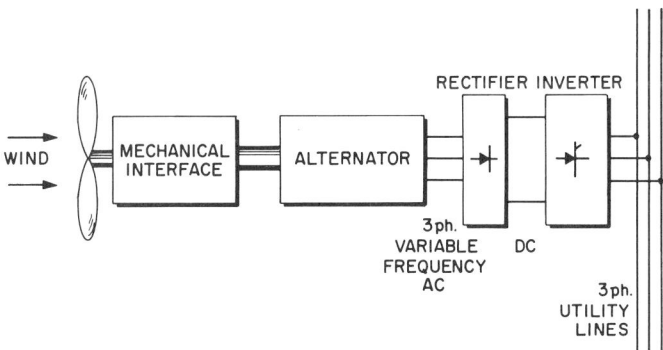

Figure 6. Schematic of Wind-Electric System Employing AC-DC-AC Link.

the desired low frequency instead of with conventional direct-current. The technique of field modulation has been proposed by several researchers in the past [18,19]. However, all the schemes suggested suffer from shortcomings of one form or another, necessitating derating of the machine to allow for additional losses that occur due to the presence of harmonics.

In the scheme developed at Oklahoma State University, a parallel-bridge rectifier system connected across the machine stator terminals yields an output which consists mainly of a full-wave rectified sine wave at the frequency of modulation; in addition there is a small ripple at six times the basic machine frequency and a small dc component. The use of parallel-bridge rectifier system eliminates line-to-line short circuits and capacitors at bridge-inputs reduce excitation requirements, improve regulation and efficiency. If the ratio of the basic electrical rotational frequency to the modulation frequency is greater than ten, the ripple and the additional dc component become negligibly small. The required low frequency output is obtained from the parallel-bridge rectifier output by using a four-thyristor switching circuit followed by a suitable filtering arrangement. The voltage at the input to the thyristor switching circuit is essentially a full wave rectified sinusoid at the required low frequency. The thyristors switch at approximately zero voltage points of this waveform to produce the desired output. Figure 7 shows a block diagram of the field modulated generator system.

A simplified single-phase schematic of the field modulated generator system is shown in Figure 8. It combines a high-reactance high-frequency (approximately 1 kHz) three-phase alternator and suitable electronic switching and output processing elements. The rotor of the alternator is laminated to handle ac excitation and the rotor winding leads are brought out through a conventional slip-ring arrangment. (It is possible to eliminate the slip rings by introducing an additional air-gap as in the case of certain types of eddy-current couplings). Both cylindrical rotor and salient pole configurations are suitable for field modulated power generation.

A parallel-bridge rectifier system consisting of three full-wave bridges, one across each of the three stator windings with their outputs tied in parallel is used at the machine terminals. The output of this rectifier system is fed to the main switching circuit formed by the four thyristors SCR1 through SCR4. The commutating circuit consisting of SCR5, SCR6, L_2 and C_2 aids in turning off two of the four thyristors at the end of each half cycle of the output. Filtering is performed by the tuned circuit consisting of filter capacitor

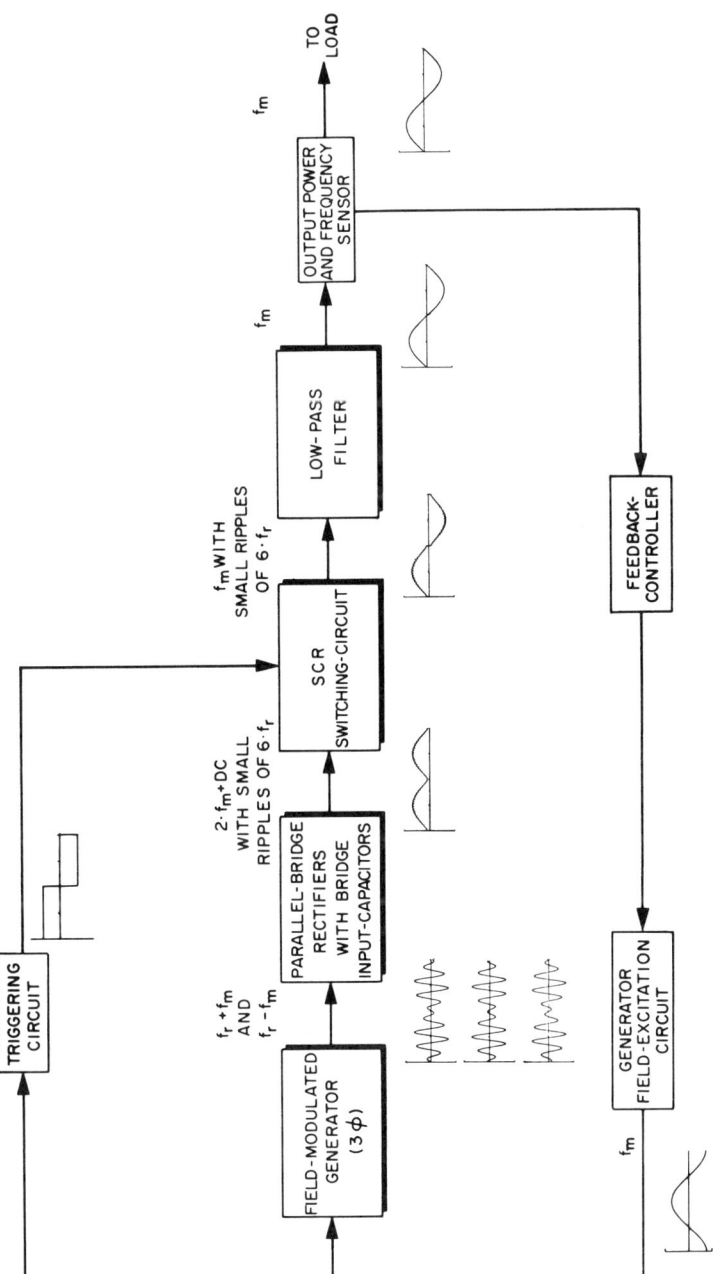

Figure 7. Block Diagram of the Field Modulated Generator System.

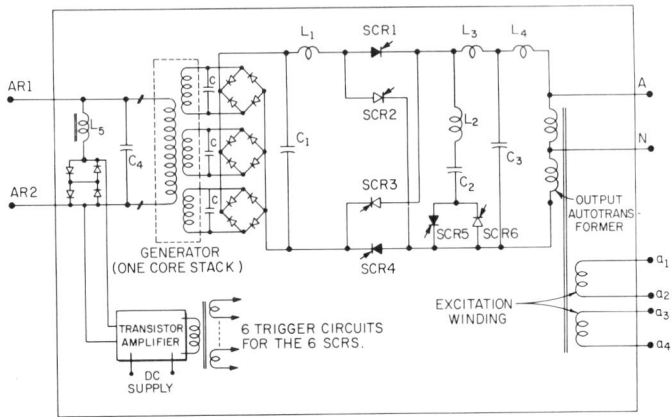

Figure 8. Simplified Single-Phase Schematic of the FMGS.

C_3 and output autotransformer.

Generation of (two) three-phase power requires (two) three separate core stacks (magnetic circuits) on a single shaft, each modulated separately while maintaining the proper phase difference between corresponding single-phase outputs. The electronics associated with each phase will be identical to the single-phase system discussed above. It can be seen that a single-phase load on a polyphase (two core stack or three core stack) field modulated generator system becomes a three-phase load on one of the core stacks.

C. Double Output Induction Generator [20]

When the rotor of an induction motor (connected to utility grid) is driven at speeds above synchronous, the machine becomes a generator and delivers constant line-frequency power to the utility grid. Since induction machines are normally designed to operate at small slips (less than 5%), rated output conditions as a generator are obtained at speeds slightly above (less than 105%) synchronous. Therefore, wind-electric systems employing conventional induction generators must operate at esentially constant speeds.

One approach to overcome this restriction and make the

induction generator a variable-speed constant-frequency device is to employ rotor power regeneration as illustrated in Figure 9. The squirrel cage rotor is replaced by a wound rotor with leads brought out through slip rings. Rotor power output at slip frequency is converted to line-frequency power by an AC-DC-AC link consisting of a rectifier and inverter as shown.

Output power is obtained both from stator and rotor and hence this device is named double output induction generator. Rotor output power has the electrical equivalence of an additional impedance in the rotor circuit. Therefore, increasing rotor outputs lead to increasing slips and higher speeds. Such an operation broadens the operating speed range from synchronous above to twice the synchronous speed. This corresponds to a per unit slip range of 0 to -1.0.

Summing up, double output induction generator is the result of combining two well developed technologies (induction machine technology and AC-DC-AC technology) to device a variable-speed constant-frequency generating system.

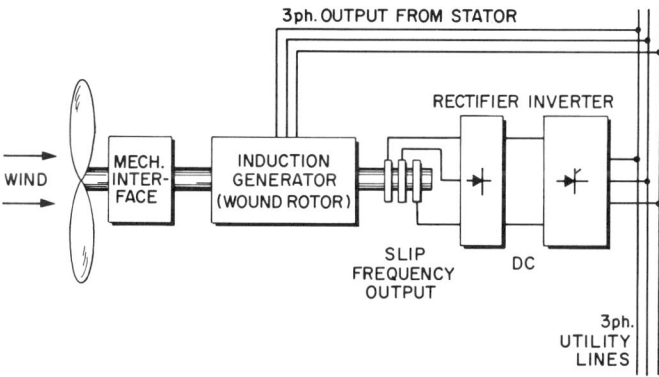

Figure 9. Double Output Induction Generator Schematic.

IV. CONSTANT SPEED SYSTEMS

A. General

Primary functions of the controller for a wind-electric conversion system are enumerated below.

(i) Start-up of the unit (if necessary as in the case of Darrieus aeroturbines).
(ii) Synchronize the unit properly (if necessary as in the case of wind-driven synchronous generators).
(iii) Monitor the wind direction and orient the aeroturbine into the wind (applicable only in the case of horizontal axis units).
(iv) Monitor the wind speed and electrical power output and load the system by properly adjusting the pitch angle of blades and generator excitation.
(v) Maintain the electrical power output at the rated value for all wind speeds above rated value by appropriate blade pitch control.
(vi) Shut down the unit under abnormal operating conditions such as
    a) prolonged motoring
    b) too high a generator temperature
    c) excessive wind speed
and d) electrical faults in the system.

B. Wind Driven Synchronous Generator

Most of the major developmental efforts in wind-electric conversion have employed synchronous generators as the logical choice. The historic Smith-Putnam 1.25 MW unit [21] on Grandpa's Knob near Rutland, Vermont (1941-1945) and ERDA/NASA's 100 kW installation [22] at Plum Brook near Sandusky, Ohio (1975) are two prime examples of this choice. The baseline design used in the wind energy mission analysis (Reference 10) employs 1800 RPM 1.5 MW 4160 V three-phase synchronous generator, equipped with an exciter and voltage regulator. Hutter's group in Stuttgart also employed synchronous generators [23] in their 100 kW experiments.

With conventional synchronous machines operating in parallel with power lines, assuming the excitation to be constant, an increase in primemover (aeroturbine) torque will tend to speed up the rotor. This will automatically increase the torque angle δ and the power delivered to establish a new equilibrium state with speed still remaining synchronous, as long as pull-out conditions are not reached. If the excitation is changed with prime-mover torque remaining constant, both the induced voltage and torque angle change to maintain the power delivered constant. Ideally, input torque controls the power delivered and excitation current controls the operating power factor.

Dynamic interaction of wind driven synchronous machines and utility networks is complicated by continuous variations in the input. System inertia has a tendency to smooth out these variations to some extent; however, influence of the coupling (Reference 13) is very predominant and an inappropriate choice may even lead to unstable operation [24]. Either a compliant shaft with damping or a rate coupling (fluid drive or eddy current coupling) should be interposed between the aeroturbine and the synchronous generator to mitigate this problem. To prevent overloads due to sharp wind gusts and to limit the output of the generator to rated value, very high performance is needed in the blade pitch control loop. Rate couplings with a limit on maximum torque (usually less than 200% of rated) turn out the smoothest performance.

During the oscillatory behavior of a synchronous machine subsequent to the occurrence of a sharp wind gust, the machine may go into motoring. During motoring, energy flows from the grid to the system and this can only aggravate the oscillations. Motoring must be avoided as far as possible and rate couplings help to accomplish this objective.

In addition to random wind gusts, wind-electric systems are subject to cyclic load variations of mechanical and aerodynamic origin. Another problem is the unavailability of substations nearby, thus necessitating the use of long distribution feeders. Studies on the dynamics of wind generator systems for large [25] and small [26] disturbances are recently gaining momentum and considerable amount of additional work is indicated by these initial efforts.

Synchronizing a wind driven generator with utility lines is a non-trivial problem in itself. Computer simulation studies have shown [27] that random synchronization can be accomplished with limited transients by proper application of an effective speed control system and an automatic synchronizer. Results of this study have been verified experimentally with the ERDA/NASA 100 kW system at Plum Brook.

In general, excitation control improves the stability of a wind driven synchronous generator operating in parallel with a power grid. A wealth of knowledge is available in the literature on improving the stability by excitation control. The techniques and the results are quite familiar to researchers in this area of power engineering.

C. Wind Driven Induction Generator [28]

Induction generators are basically simpler than synchronous generators. They are easier to operate, control and maintain, have no synchronization problems and are economical. Over the years, they have logged several years of operating experience with aeroturbines; the most notable of which being the Danish Gedser mill (1957). However, they draw their excitation from the grid and consequently impose a reactive voltampere burden. To correct this situation, either static capacitors or some kind of a rotor feeding scheme is necessary. In addition, efficiencies of induction generators are slightly lower than the efficiencies of synchronous generators over the entire operating range.

Wind driven induction generators do not require a rate coupling or a compliant shaft with damping to perform stably in the presence of wind gusts. Output capability must still be limited to the rated value by blade pitch control. Electrical output is uniquely determined by the operating speed and once the output exceeds the rated value, pitch control should come into action, reduce C_p and limit the output. Sudden disturbances such as wind gusts are taken care of by load changes. Since the output increases as the speed increases (up to the pull-out value), overspeed protection is very vital for the safety of the generator and the overall system.

In operation and control, induction generators are far simpler than synchronous generators. Even a direct short across the lines is not very critical since the excitation to the induction generator will then go to zero automatically.

It is attractive to use the same machine as a motor for starting the aeroturbine (when it is not self starting) and as a generator after attaining the proper speed. Although such an arrangement has been used with smaller units, two separate machines are preferable in the case of large wind-electric systems because of the different characteristics (mainly torque-speed) required for the two modes of operation.

V. VARIABLE SPEED SYSTEMS

A. General

The fundamental requirement of any control scheme for variable-speed constant-frequency wind energy systems is that it allows the system to deliver the right amount of electrical power needed to maintain a constant tip-speed ratio. The general nature of variation of aeroturbine angular speed, shaft torque and mechanical power output for variable speed systems is illustrated in Figure 10. The control scheme should adjust the power delivered such that these ideal operating conditions are approached as closely as possible.

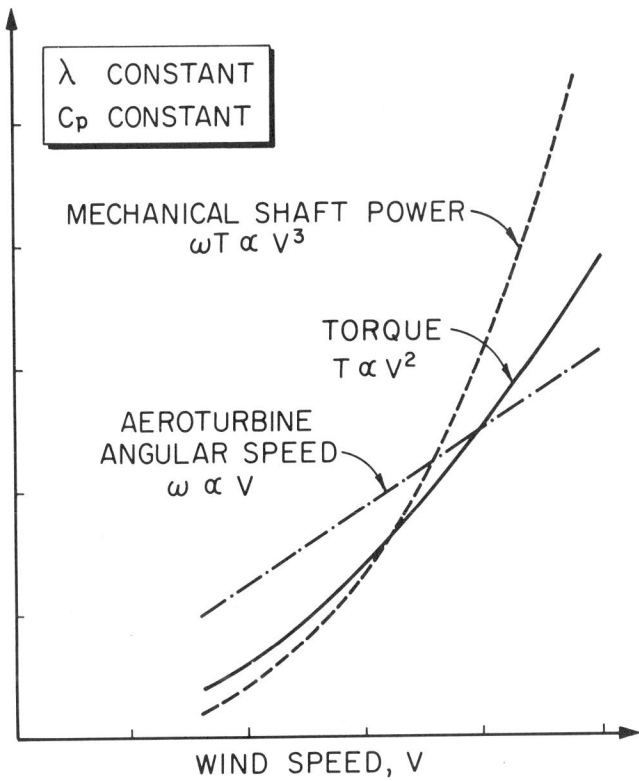

Figure 10. Illustrating Ideal Operating Conditions for Variable-Speed Wind-Electric Systems.

B. Control Philosophy [29]

Maintenance of a constant tip speed ratio in the purest sense means an instantaneous change in shaft speed for an instantaneous change in wind speed (wind gust). Due to the natural time constant of the system, instantaneous response to wind gusts would not be possible without some action by the controller. With no control, a wind gust would cause an instantaneous decrease in the tip speed ratio, and therefore C_p.

If C_p decreased enough, it is possible that the uncontrolled system would be unable to respond at all, and the additional wind energy would be spilled by the aeroturbine. If it did respond, it would be extremely slow due to the large mechanical time constants and the counter-torque produced by the increasing load.

In order for the system to respond to wind gusts, the controller would have to cause the shaft speed to respond as quickly as possible to the wind. This would require a reduction in loading to allow faster shaft response and would still not be perfect due to the mechanical time constant. In view of the above discussions, it is felt that control of the system to respond to wind gusts of certain short time durations is impractical for the following reasons.

(i) The controller would have to be more complex and, therefore, more expensive.

(ii) The additional energy to be gained by allowing the system to respond to wind gusts would be at least partially offset by energy losses due to reduction in loading to decrease system response time.

(iii) In some areas the system would constantly be in a transient condition due to the high frequency of wind gusts.

As pointed out earlier, with proper monitoring and control, the VSCF system may be allowed to deliver additional energy at wind speeds in excess of the rated wind speed. The limiting factor in this case is machine dissipation and the associated temperature rise. If the system is designed to deliver its rated output power at the rated wind speed V_R, then by monitoring machine temperature, the system may be allowed to deliver more than its rated power for wind speeds in excess of V_R up to 1.5 or 2 V_R. At higher wind speeds the machine cooling would be more efficient, and the generator would be able to deliver more than its rated power for a certain time duration. The controller would simply monitor generator temperature and cause a decrease in output when it reaches a predetermined maximum safe value.

C. Wind Driven FMGS [30]

1. General

Figure 11 shows a simplified model of a field modulated generator system pumping constant frequency power into a power grid at a certain wind speed. For the FMGS, the model is

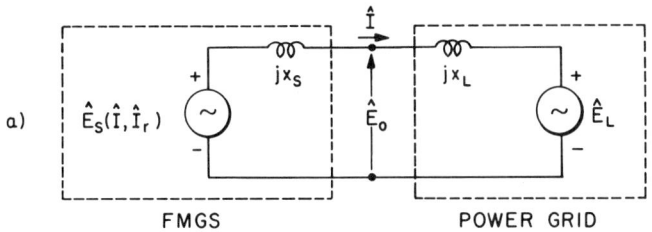

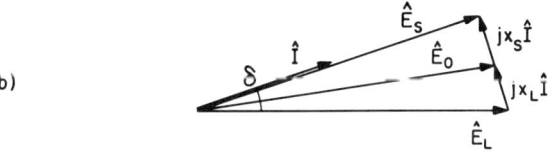

Figure 11. Circuit Model of FMGS Pumping Power into a Power Grid.

essentially an ideal voltage source $\hat{E}_s$, which is dependent on the output current $\hat{I}$ and the rotor excitation current $\hat{I}_r$. The series reactance X_s accounts for the output transformer which is also a part of the output filter of the FMGS. The power grid at the input point is modeled in the conventional way using an ideal voltage source $\hat{E}_L$ in series with a reactance X_L.

The phase angle between the two ideal voltage sources, $\hat{E}_s$ representing the generator (at the output of the inverter thyristor) and $\hat{E}_L$ representing the power grid is not affected by changes in prime-mover (aeroturbine) torque. It can be controlled externally by advancing or retarding the phase of rotor applied voltage with respect to grid voltage. An increase in torque input will speed up the system with resulting increase in the magnitude of $\hat{E}_s$ without a change in δ.

It is prudent to operate wind energy systems pumping power into power lines in a manner that is optimum from the viewpoint of the wind system. For wind driven FMGS, this translates into maintaining $\hat{E}_s$ and $\hat{I}$ in-phase as shown in Figure 11. Assuming the system to be operating in this manner, an increase in the magnitude of $\hat{E}_s$ without a change in δ will alter operating conditions in two ways: 1) the magnitude of $\hat{I}$ will increase and 2) $\hat{I}$ will lag $\hat{E}_s$. The net result of these two changes will be to increase the power generated and slightly decrease the efficiency of the system. A suitable increase in δ will, however, bring $\hat{E}_s$ and $\hat{I}$ back in phase and alter the power output to match the new operating conditions. If excitation current magnitude is increased with input torque constant, the system will slow down to a new value of speed at which input-output power balance is maintained with the increased excitation current and the new efficiency value.

From the above discussion, it appears that power delivered by a wind driven FMGS to a power grid can be controlled by controlling the magnitude and phase of excitation voltage applied to rotor terminals. A scheme to accomplish this is shown in Figure 12.

2. Control Schemes

Two schemes to integrate wind driven field modulated generator systems with conventional utility grids are presented in this section. One of them uses wind speed and aeroturbine shaft speed as control inputs and the other uses aeroturbine shaft torque and speed as control inputs.

    a. Wind speed as a control input. Monitoring wind speed

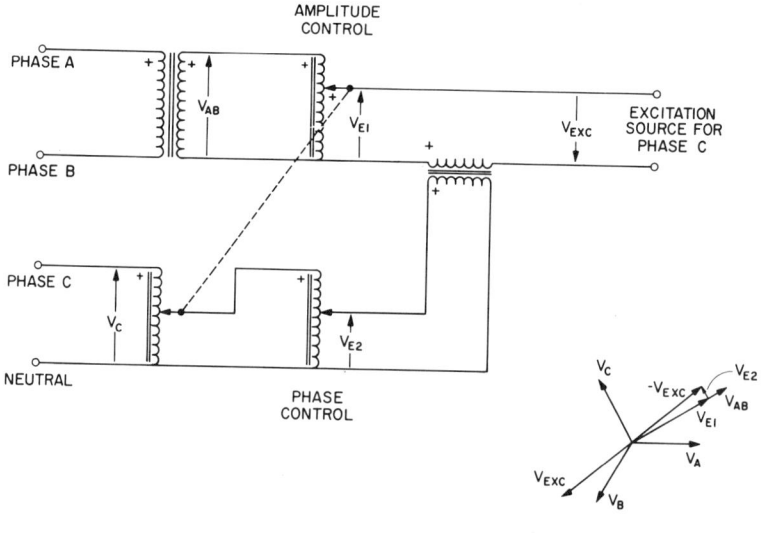

Figure 12. A Scheme to Vary FMGS Excitation Voltage.

and using it to control the electrical output appears to be a logical way to maintain a constant tip speed ratio with varying wind speeds. The electrical output must be adjusted such that the aeroturbine shaft speed changes proportionally with the wind speed. Figure 13 shows a simplified block diagram of the control scheme proposed. Whenever shaft speed and wind speed are not proportional, an error signal is developed and this signal initiates the necessary changes (both amplitude and phase) in the (excitation) voltage applied to the rotor of the FMGS to adjust the electrical power output.

Output of the wind-speed sensor is filtered to allow the controller to see only average wind speed changes and not respond to wind gusts of short durations. The major difficulty in using wind speed as a control input is deciding where to monitor the wind. Depending on the type of aeroturbine used, the air surrounding the installation will be quite turbulent. The wind speed sensor would then have to be located at some suitable distance from the aeroturbine. There will be some difference (in magnitude, direction and time) between the wind speeds at the sensor and turbine locations. Additionally, the sensor system should be designed to track wind direction identically (yaw control) with the aeroturbine. It is believed that these errors may be no more serious than similar errors existing in other more complicated control schemes.

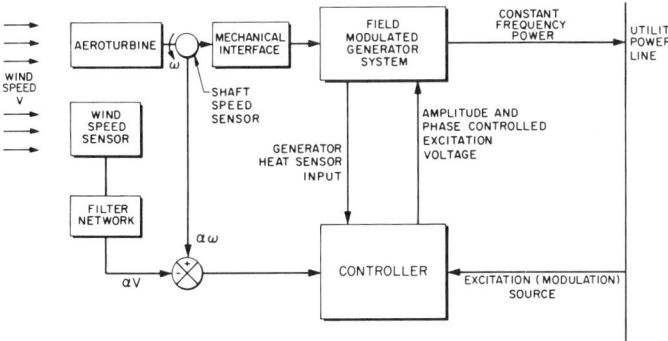

Figure 13. Block Diagram of Control Scheme with Wind Speed and Shaft Speed as Control Inputs.

b. Torque as a control input. One way to overcome the problem associated with location of wind speed sensor is to use aeroturbine shaft torque rather than wind speed as a control input. Changes in wind speed are instantaneously translated to changes in shaft torque. Therefore, for reasons discussed earlier, output of the torque sensor must be filtered to allow the controller to see only average changes and not respond to short-term wind gusts.

Optimum performance requires the shaft speed to vary proportionally to wind speed and shaft-torque to vary proportionally to the square of the wind speed. Therefore, the filtered torque signal can be compared with the squared shaft speed signal to generate an error signal to actuate the controller. This scheme is illustrated in the simplified block diagram shown in Figure 14. Alternately, a square root circuit may be used at the output of the filter to obtain a signal from torque sensor, but proportional to wind speed. This can then be compared with a signal proportional to shaft speed (without the squarer) to generate the error signal. Either way, the shaft speed must change by the square root of the factor by which torque changes.

One problem in using torque as a control input must be

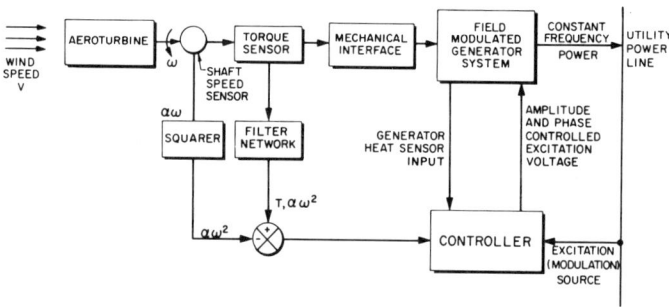

Figure 14. Block Diagram of Control Scheme with Torque and Shaft Speed as Control Inputs.

brought to light. If the coefficient of performance of the aeroturbine changes rapidly with changes in tip speed ratio (highly peaked C_p vs λ curve), then an increase in wind speed may instantaneously result in a lower torque instead of a higher torque. This may lead to misoperation of the control system. Smoothing action inherent in the mechanical system, proper design of the filter network and an aeroturbine with a moderate variation in C_p vs λ may be combined suitably to eliminate this problem altogether.

The controllers used in both schemes have an additional input proportional to generator temperature. If, due to some reason such as continued overloading or fault, the temperature of the machine goes above some safe predetermined value, the controller would decrease the electrical output or even disconnect the system completely from the utility network as the situation warrants.

D. AC-DC-AC Link

The technology of obtaining variable-speed constant-frequency via an AC-DC-AC link is very similar to that of high voltage DC systems. Considerable literature exists on this topic [31]. These converters belong to the naturally commutated (or line commutated) group. They utilize an ac source which periodically reverses polarity and causes commutation to occur naturally. Since frequency is automatically fixed by the line, they are also known as 'synchronous inverters'. Control of such converters is achieved by delaying the firing angle of the thyristors involved by building a controllable phase delay into the switching control signal generator [32].

E. Double Output Induction Generator

In this case, control of output power is obtained by varying the firing angle α of the line-commutated inverter in the regeneration circuit. The equivalent rotor circuit resistance as seen from the stator side changes with variations in α and this, in turn, changes the rotor speed, output from stator and the output from rotor.

The total electrical power output from this system is equal to the sum of the outputs from stator and rotor. Therefore, theoretically it is possible to obtain twice the rated power output (rated as a regular induction generator with short-circuited rotor) by operating the system at a per unit slip of -1.0 corresponding to the rotor turning at twice the synchronous speed. The rotor induced voltage under these conditions will be equal to the value under rotor blocked conditions as an induction motor. Assuming ideal conditions, both rotor and stator currents will be 1.0 per unit and therefore, without exceeding the voltage and current ratings of stator and rotor, output of the machine could be doubled. However, such an operation would require the slip rings to be capable of handling rated power.

In short, the controller should program the inverter firing angle α to match the total electrical output with the wind input while maintaining constant tip speed ratio for the aeroturbine under varying input wind conditions.

VI. CONCLUDING REMARKS

Operation and control of wind-electric conversion systems pose several challenging problems, many of which are yet to be studied in detail. Although it is a fairly simple matter to put a system together and obtain 'some' electrical energy output, the problem of optimizing the energy obtained from an aeroturbine of a certain diameter located in a certain wind regime is indeed very complex. Optimization of the energy output is quite important because of the capital intensive nature of wind-electric systems.

The options available in aeroturbine design [33](fabrication, structure, control, location and speed), mechanical interface and generator type are many and numerous combinations of these appear to be viable both from the technical and economic view points. Depending on the size of the aeroturbine, certain factors appear to dominate the design choice. For example, in the case of very large diameter (40 to 50 m and above) units, preserving mechanical stability may outweigh all other considerations. With small and medium size systems, there may be more room to maneuver towards achieving optimum energy collection.

At present, constant speed approach employing synchronous generators is getting considerable attention. In this connection, further work is indicated in several areas, a few of which are mentioned below.

(i) Optimum choice of coupling and its characteristics; in the case of eddy current couplings, it is possible to utilize the dependence of the torque-slip characteristic on the excitation current (see Figure 15) to perform a major control function.

(ii) Utilizing short-term (less than one minute) energy storage to smooth out the output of a wind-electric system.

(iii) Optimum choice of locations based on statistical analysis of wind regimes.

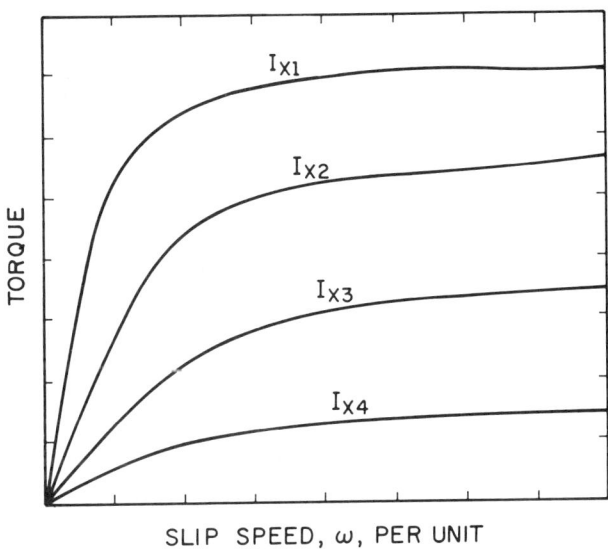

Figure 15. Illustrating the Nature of Eddy Current Coupling Characteristics.

(iv) Since wind is random by nature, it should be possible to apply the wealth of knowledge available in stochastic control and systems theory to improve the design of wind-electric conversion systems.
(v) Dynamics of multi-unit large wind-electric arrays interfaced with utility grids.
(vi) Systematic studies on the influence of inertia, electrical damping, system reactance, machine reactance and various types of voltage control on the operation and stability of a wind-electric system.
(vii) Use of microprocessors and telemetry to control a group of wind-electric systems located in a geographical region.

Wind-electric systems employing induction generators should be investigated further, especially in the areas of dynamics and stability of multi-unit arrays, self excitation, control and protection.

Variable-speed wind-electric systems appear to be most attractive in the 'small' (10 to 50 kW) and '100 kW scale' (50 to 250 kW) sizes and for use in large-capacity 'multirotor on one tower' concepts. Loosening the severe speed restrictions imposed by synchronous and induction generators and allowing the speed to vary by, say ± 10%, may prove to be beneficial even in the case of large wind-electric systems. This prospect must be explored further. Obviously, some kind of a variable-speed constant-frequency generating system will then be necessary.

Influence of unidirectional power flow (non-motoring behavior) on the stability of multi-unit arrays should be studied. A beneficial impact, coupled with the possibility of allowing ± 10% speed changes, would make a strong case for replacing conventional synchronous and induction generators by other generation schemes. The possibility of loading the generator above its rated value at high wind speeds with different generation schemes should also be explored.

Efficiencies of electrical generating equipment decrease at the lower end of their operating range. This decreases the overall efficiencies of wind-electric systems at lower wind speeds. If efficiencies can be improved in this range, plant (load) factors can be increased, leading to a decrease in the cost of wind generated electrical energy. With wind driven field modulated generators, the system can be switched to the single-phase mode of operation so as to derive the entire electrical output from a single core stack and realize higher mechanical-to-electrical efficiencies. At higher wind speeds, operation can be reverted to two or three-phase mode. Such a procedure will maintain a high overall efficiency for the wind-electric system during most of the operating range and result in higher plant factors. Several wind-driven FMGS can operate in parallel and employ one common inverter for output conditioning and interfacing with the utility grid. Paralleling will be done at the outputs of the parallel-bridge rectifier systems (see Figure 8) associated with the individual generators. A schematic of this approach is shown in Figure 16. Outputs of a cluster of aeroturbine-generator units located in a windmill farm can thus be processed through one inverter-transformer station. It is also possible to employ this 'common inverter' concept in a 'multi-rotor on one tower' approach being suggested to build large (in capacity) wind energy conversion systems and take advantage of the economy of mass production of standardized smaller (about 6 to 18 m diameter) units. All these aspects require further study and development [34].

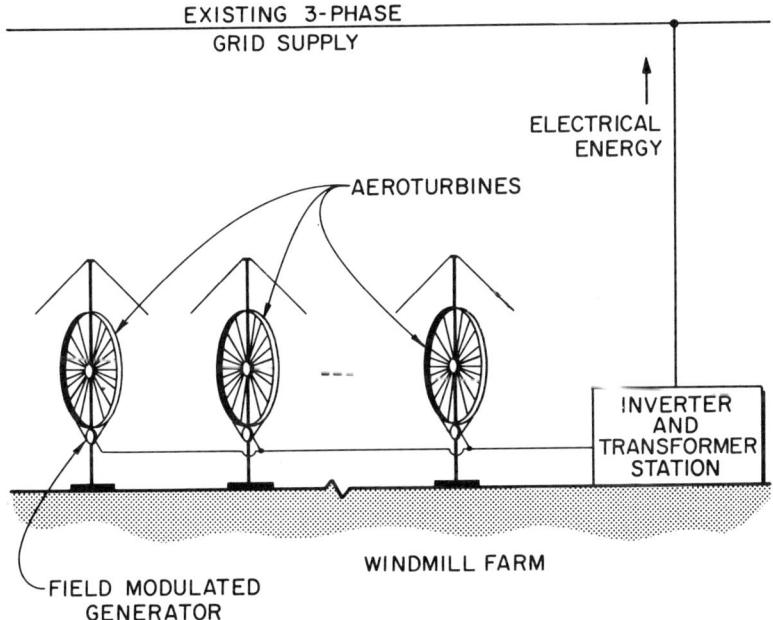

Figure 16. Schematic of the Common Inverter Concept.

Design of the control system for variable speed wind-electric systems will involve decisions regarding the minimum duration of wind gust for which the system should respond, circuitry to be used to filter information concerning wind input and refinements in the control philosophy and strategy.

Utilization of wind power is not just a thing of the past; it is taken very seriously [35] at present and wind-electric conversion is expected to lead the list of possible uses. Only future research will determine what will be the most 'cost effective' wind system for different applications, wind regimes and sizes. Meanwhile, the engineers and scientists involved in the development of this renewable energy resource have many challenges to face and problems to solve.

VII. REFERENCES

(1). NSF/NASA Solar Energy Panel Report,"An Assessment of Solar Energy as a National Energy Resource", 1972.

(2). Savino, J. M., Editor. "Wind Energy Conversion Systems", Proc. Workshop held in Washington, D. C., No. NSF/RA/W-73-006, 1973.

(3). Eldridge, F. R., Editor. "Wind Workshop[2]", Proc. Second Workshop on Wind Energy Conversion Systems, No. NSF-RA-N-75-050, MTR-6970, 1975.

(4). Ljungström, O., Editor. "Advanced Wind Energy Systems", Proc. Workshop held in Stockholm, Sweden, 1976.

(5). Reynolds, J., "Windmills and Watermills", p. 6-13. Praeger Publishers, New York, 1970.

(6). Putnam, P. C., "Power from the Wind", Van Nostrand, New York, 1948 Republished 1975.

(7). Golding, E. W., "Generation of Electricity by Wind Power", E. & N. Spoon, Ltd., 1955.

(8). Ramakumar, R. et al., "Prospects for Tapping Solar Energy on a Large Scale", Solar Energy 16(2), 107, 1974.

(9). Ramakumar, R. et al., "Wind Energy Utilization Prospects", in Proc. 21st Annual Meeting of the Institute of Environmental Sciences, Vol. 1 (Energy and the Environment), pp. 138-142, 1975.

(10). General Electric Company Final Report, "Wind Energy Mission Analysis", No. COO/2578-1/1-3, prepared for the United States Energy Research and Development Administration, 1977.

(11). Ramakumar, R., "Wind-Electric Conversion Utilizing Field Modulated Generator Systems", in Proc. Sharing the Sun! Solar Technology in the Seventies Conf., Vol. 7, pp. 215-229, Winnipeg, Manitoba, Canada, 1976 ; to appear in Solar Energy.

(12). Thresher, R. W. and Wilson, R. W., "Design Considerations for the Darrieus Rotor", in Proc. 11th Inter. Energy Conv. Engg. Conf., Vol. 2, pp. 1787-1794, State Line, Nevada, 1976.

(13). Johnson, C. C. and Smith, R. T., "Dynamics of Wind Generators on Electric Utility Networks", IEEE Trans. Aerospace and Electronic Systems AES-12 (4), 483, 1976.

(14). Glazenko, T. A., "Some Problems in the Design of an Asynchronous Clutch with a Monolithic Rotor", Avtomatika i Telemekhanika 19(8), 1958; English Translation, Automation and Remote Control, pp. 783-790, 1959.

(15). Ramakumar, R. et al., "Description and Performance of a Field Modulated Frequency Down Converter", in 1972 SWIEEECO Record, IEEE Catalog No. 72 CHO 595-9 SWIECO, pp. 252-256, 1972.

(16). Allison, H. J. et al., "A Field Modulated Frequency Down Conversion Power System", IEEE Trans. Industry Applications IA-9 (2), 220, 1973.

(17). Tsung, C. C., "A Study of the Field Modulated Generator

System", Ph.D. Thesis, Oklahoma State University, Stillwater, Oklahoma, 1976.
(18). Wickson, A. K., "A Simple Variable Speed Independent Frequency Generator", AIEE Conference Paper No. CP 59-915, Summer and Pacific General Meeting, Seattle, Washington, 1959.
(19). Bernstein, T. and Schmitz, N. L., "Variable Speed Constant Frequency Generator Circuit Using a Controlled Rectifier Power Demodulator", AIEE Conference Paper No. CP 60-1053, Pacific General Meeting, San Diego, California, 1960.
(20). Yadavalli, S. R. and Jayadev, T. S., "A New Generation Scheme for Large Wind Energy Conversion Systems", in Proc. 11th Inter. Energy Conv. Engg. Conf., Vol. 2, pp. 1761-1765, State Line, Nevada, 1976.
(21). Savino, J. M., "A Brief Summary of the Attempts to Develop Large Wind-Electric Generating Systems in the U.S.", NASA Technical Memorandum No. TMX-71605, 1974.
(22). Thomas, R. et al., "Plans and Status of the NASA-LEWIS Research Center Wind Energy Project", NASA Technical Memorandum No. TMX-71701, 1975.
(23). "Betriebserfahrungen Mit Der 100 kW Windkraftanlage Der Studiengesellschaft Windkraft E.V. Stuttgart", Sonderdruck Aus Brennstoff-Wärme-Kraft, Vol. 16 (7), pp. 333-340, 1964.
(24). Qazi, A. Q. and Ramakumar, R., "Behavior of Wind Driven Synchronous Generators Under Wind Gusts", in 1977 Control of Power Systems Conference Record, IEEE Catalog No. 77 CH 1168-4 REG5, pp. 20-25, College Station, Texas, 1977.
(25). Romanelli, P. J., "Electrical Generating Equipment and Electric Utility Requirements for High-Power Wind Generator Systems", in Record of the Tenth Inter. Energy Conv. Engg. Conference, IEEE Catalog No. 75 CHO 983-7 TAB, pp. 1251-1257, Newark, Delaware, 1975.
(26). Pantalone, D. K. and Potter, A. G., "Application of Wind Power to the Electric Power System", in Proc. Frontiers of Power Technology Conference, pp. 6-1 to 6-22, Stillwater, Oklahoma, 1976.
(27). Hwang, H. H. and Gilbert, L. J., "Random Synchronization of the ERDA-NASA 100 kW Wind Turbine Generator with Large Utility Networks", in 1977 Control of Power Systems Conference Record, IEEE Catalog No. 77 CH 1168-4 REG5, pp. 26-30, College Station, Texas, 1977.
(28). Jayadev, T. S., "Induction Generators for Wind Energy Conversion Systems". Report submitted to U.S. ERDA; AER 75-00653, Feb. 1976.
(29). Ramakumar, R. and Ashbaugh, C. D., "Operational and

Control Aspects of Wind Energy Systems Utilizing Field Modulated Generators", in Record of the 1976 Control of Power Systems Conference, IEEE Catalog No. 76 CH 1057-9 REG5, pp. 111-114, Oklahoma City, Oklahoma, 1976 .

(30). Ramakumar, R., "Wind Driven Field Modulated Generator Systems", in Proc. 11th Inter. Energy Conv. Engg. Conf., Vol. 2, pp. 1766-1772, State Line, Nevada, 1976 .

(31). Kimbark, E. W., "Direct Current Transmission", Vol. 1, Wiley-Interscience, 1971 .

(32). Reitan, D. K., "A Progress Report on Employing a Non-Synchronous AC/DC/AC Link in a Wind Power Application", in Proc. 2nd Wind Workshop, pp. 290-297, Washington, D. C., 1975 .

(33). Lindley, C. A., "Promises and Problems of Wind Electric Generation", Presented at the First Helioscience Institute, Palm Springs, California, 1976 .

(34). Ramakumar, R. et al., "Development and Adaptation of Field Modulated Generator Systems for Wind Energy Applications", Final Report, submitted to ERDA Division of Solar Energy, Contract No. ERDA/NSF-AER 75-00647, 1976 .

(35). Divone, L. V. and Savino, J. M., "The US - NSF/NASA Wind Energy Conversion System Program", Proc. Advanced Wind Energy Systems Workshop (1974), Vol. 2, pp. 7-25 to 7-33, Stockholm, Sweden, 1976 .